FORTRAN 77
for Engineers

Delores M. Etter
Department of Electrical and Computer Engineering
University of Colorado, Boulder

The Benjamin/Cummings Publishing Company, Inc.

Redwood City, California · Menlo Park, California
Reading, Massachusetts · New York · Don Mills, Ontario
Wokingham, U.K. · Amsterdam · Bonn · Singapore
Tokyo · Madrid · San Juan

This is a module in *The Engineer's Toolkit,* a Benjamin/Cummings Select edition. Contact your sales representative for more information.

ISBN 0-8053-6453-6

The Benjamin/Cummings Publishing Company, Inc.
390 Bridge Parkway
Redwood City, California 94065

Contents

1 Solving Problems with FORTRAN 77

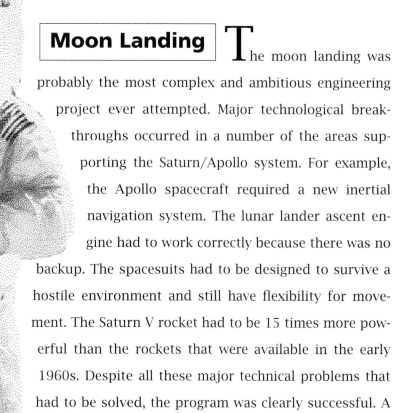

Moon Landing The moon landing was probably the most complex and ambitious engineering project ever attempted. Major technological breakthroughs occurred in a number of the areas supporting the Saturn/Apollo system. For example, the Apollo spacecraft required a new inertial navigation system. The lunar lander ascent engine had to work correctly because there was no backup. The spacesuits had to be designed to survive a hostile environment and still have flexibility for movement. The Saturn V rocket had to be 15 times more powerful than the rockets that were available in the early 1960s. Despite all these major technical problems that had to be solved, the program was clearly successful. A total of 6 spacecraft and 12 astronauts landed on the moon between 1969 and 1972.

INTRODUCTION

This chapter introduces you to problem solving by first examining the class of problems FORTRAN 77 was designed to solve. The chapter then presents a five-step process that can be applied to solving a wide variety of problems and gives a simple example that uses this five-step process to compute the average of a set of laboratory measurements.

1-1 FORTRAN 77

A *computer* is a machine designed to perform operations that are specified with a set of instructions called a *program.* Computer *hardware* refers to the computer equipment, and computer *software* refers to the programs that describe the steps we want the computer to perform. This software can be written in a variety of languages and for a variety of purposes. This section describes the FORTRAN 77 language; the next section discusses the problem-solving process we will use to write FORTRAN 77 programs.

High-level languages are computer languages that have English-like commands and instructions. Originally developed in the 1950s for technical applications, FORTRAN (short for FORmula TRANslation) was one of the first high-level computer languages. Since FORTRAN's creation, its power, speed, and ease of use have been consistently enhanced with new features. Today FORTRAN is a widely used programming language for solving scientific and engineering problems because of its ability to perform mathematical calculations quickly and efficiently. Another programming language you may encounter in your career is C, which is also frequently used to solve engineering and scientific problems.

Alternatives to writing a program in a high-level language exist, such as using specialized software tools and applications packages. Prewritten software routines designed to solve common scientific and engineering problems allow you to avoid writing a program that has already been created. Or you can use the powerful engineering, mathematical, and statistical functions available in spreadsheets, such as Lotus 1-2-3 and Quattro Pro. For even more specialized calculations, you can use mathematical packages, such as MathCAD and MATLAB. However, when pre-existing software tools do not exist, or you don't have the time to learn how to use an applications package, or the problem you're working on is too large or complex for an applications package, writing a program in a high-level language is the best strategy.

FORTRAN 77 is a specific version of FORTRAN based on standards established in 1977. This standard greatly improved the language by adding new features and structures that allow you to write powerful yet readable programs. A new standard, *Fortran 90*, provides a number of very powerful additions to the language; Appendix B summarizes the features in Fortran 90.

Because we have been trained to think in terms of English-like phrases and formulas, we prefer to use high-level languages such as FORTRAN 77 to tell the computer the steps we want it to perform. A special program called a *compiler* is then needed to translate a FORTRAN 77 program into a language that the computer can understand called machine language.

Learning a high-level language is similar to learning a foreign language. Each step you want the computer to perform must be described in the specific computer language *syntax,* or grammar. Fortunately, computer languages have small vocabularies and no verb conjugations; however, computers are unforgiving in punctuation and spelling. A comma or letter incorrectly placed will cause errors that keep your program from working properly.

1-2 A FIVE-STEP PROBLEM-SOLVING PROCESS

Problem solving is an activity in which we participate every day. Problems range from analyzing chemistry lab data to finding the quickest route to work. Computers can solve many of our problems if we learn how to communicate with them in computer languages such as FORTRAN 77. Some people believe that if they describe a problem to a computer, it will solve it for them. Programming would be simpler if this were the case; unfortunately, it is not. Computers can perform only the steps that you specify in detail. You may wonder then, why go to the effort of writing computer programs to solve problems if every step must be described carefully? The answer is that computers can perform the tasks extremely accurately and with fantastic speed (millions of arithmetic operations per second). In addition, computers never get bored. Imagine sitting at a desk analyzing laboratory data for 8 hours a day, 5 days a week, year after year. Yet thousands of laboratory results must be analyzed every day. Once the steps involved in performing a particular analysis (such as computing an average) have been carefully described to a computer, it can analyze data 24 hours a day, with more speed and accuracy than a group of technicians.

Much of this module will focus on teaching you the FORTRAN 77 language, but any computer language is useless unless you can break a problem into steps that a computer can perform. The approach you will use to write FORTRAN programs is based on the following five-step problem-solving procedure. A general procedure is shown on the left, and the procedure you will use to write your programs is shown on the right.

A General Problem-Solving Model	FORTRAN Problem-Solving Model
1. Define the problem.	State the problem clearly.
2. Gather information.	Describe the input and output information.
3. Generate and evaluate potential solutions.	Work the problem by hand (or with a calculator) for a simple set of data.
4. Refine and implement a solution.	Develop a solution that is general in nature.
5. Verify and test the solution.	Test the solution with a variety of data sets.

To better understand these steps, consider the familiar problem of computing an average. This task arises in many applications, such as in summarizing data from a lab experiment.

Step 1 defines the problem. You need to study the problem to determine what the program is supposed to accomplish, and then state the problem clearly. For this example the problem statement is as follows:

Compute the average of a set of experimental data values.

Step 2 consists of gathering information. You need to describe carefully any information or data needed to solve the problem and then identify the values to be computed. These items represent the input and output for the problem; collectively they are called *input/output (I/O)*. The I/O description in this example is the following:

Input—the list of experimental data values
Output—the average of the data values

Step 3 generates and evaluates potential solutions. During this step you need to work the problem by hand or with a calculator, using a simple set of data. The following example computes the average of a set of values.

Lab Measurements—4/23/95

Number	Value
1	23.43
2	37.43
3	34.91
4	28.37
5	30.62
	154.76

Average = sum/5 = 30.95

For a simple problem such as this, there may be an obvious solution, and identifying it may take only a few minutes. More complex problems may be solved in several ways, and determining the most suitable solution can be a difficult and time-consuming task.

Step 4 refines and implements a solution. In this step you describe, in general terms, the operations you performed by hand. The sequence of operations that solves the problem is called an *algorithm.* The procedure that you use in your algorithm development is called *top-down design,* because you start at the *top* with the original problem and break it *down* into smaller problems that can be addressed separately. Top-down design uses two techniques: *decomposition* and *stepwise refinement.* Decomposition is a form of "divide and conquer" in which you identify the pieces of the problem that must be solved sequentially. Stepwise refinement successively refines each smaller piece of the solution using more detail. The refining continues until the solution is specific enough to convert into computer instructions.

Top-down design has several advantages. First, it helps you write programs that are easier to understand, because the instructions in the computer program will follow the initial decomposition in a block diagram to emphasize that you are breaking the solution into a series of sequentially executed steps. To assist you in refining the decomposition into a more specific solution, you use flowcharts, which show the steps in an algorithm in graphic form, or pseudocode, which presents algorithm steps in a series of English-like statements. Chapter 3 presents the details of flowcharts and pseudocode.

Decomposition

Read the data values and sum them.
Divide the sum by the number of data values.
Print the average.

The refinement of the decomposition for this example uses pseudocode to specify the details or to outline the steps in the algorithm.

Pseudocode

1. Set the sum of the values to zero.
2. Set a count of the values to zero.
3. As long as there are more data values,
   Add the next data value to the sum.
   Add 1 to the count.
4. Divide the sum by the count to get the average.
5. Print the average.

Once the detailed algorithm is described, you are ready to convert it into FORTRAN 77. The statements for solving the problem will not all be presented until Chapter 3, so there are items in the solution that you probably won't understand at this time. You can, however, see how similar it is to the decomposition and pseudocode. A data value of zero indicates the end of the data values to the program.

FORTRAN Program

```
*------------------------------------------------------------------*
      PROGRAM COMPUT
*
*  This program computes and prints the average
*  of a set of experimental data values.
*
      INTEGER COUNT
      REAL SUM, X, AVE
*
      SUM = 0.0
      COUNT = 0
*
      READ*, X
    1 IF (X.NE.0.0) THEN
         SUM = SUM + X
         COUNT = COUNT + 1
         READ*, X
         GO TO 1
      END IF
*
```

```
      AVE = SUM/REAL(COUNT)
*

      PRINT 5, AVE
    5 FORMAT (1X,'THE AVERAGE IS ',F5.2)
*

      END
*-------------------------------------------------------------------*
```

Step 5 in the problem-solving process verifies and tests the solution. You need to test the FORTRAN solution with a variety of data sets. You first execute the program to be sure the syntax is correct. You then test the program to be sure the logic is correct, and this is more difficult than testing for correct syntax. You test the program with a number of typical sets of data to be sure it works properly. A correct algorithm to average data should work properly for any set of data. If the data from our hand example were used in the FORTRAN program presented here, the output on your terminal screen would be the following:

```
THE AVERAGE IS 30.95
```

Algorithm testing is discussed further in later chapters.

The-five step problem-solving process is demonstrated throughout this module in applications. The discipline of the applications are as follows:

Applications	Across the Disciplines	
Applications	Discipline	Chapter
Bacteria Growth	Biology	2
Stride Estimation	Mechanical Engineering	2
Light Pipes	Optical Engineering	3
Rocket Trajectory	Aerospace Engineering	3
Timber Regrowth	Environmental Engineering	3
Critical Path Analysis	Manufacturing Engineering	4
Sonar Signals	Acoustical Engineering	4
Wind Tunnels	Aerospace Engineering	5
Power Plant Data Analysis	Power Engineering	5
Oil Well Production	Petroleum Engineering	6
Simulation Data	Electrical Engineering	7
Protein Molecular Weights	Genetic Engineering	8

SUMMARY

This chapter described FORTRAN 77 as a language develope
neering and scientific problems easily and efficiently. Becau
involves problem solving, it is important to begin your in
FORTRAN with a solid methodology for solving problems an
how to use the computer to help solve the problems encoun   .. discussed the fundamental concepts of computers and computing and the
process for converting a problem solution into a form the computer can
understand and execute. The following five-step procedure for developing
problem solutions (algorithms) was presented:

1. State the problem clearly (problem statement).
2. Describe the input and output information (input/output description).
3. Work the problem by hand (or with a calculator) for a simple set of data (hand example).
4. Develop a solution that is general in nature (algorithm development).
5. Test the solution with a variety of data sets (testing).

The solution is developed using top-down design. Decomposition assists in describing the general steps that have to be performed to solve the problem. Stepwise refinement guides you in refining the steps and adding necessary details. We will use FORTRAN 77 to implement these solutions in the chapters that follow.

Key Words

algorithm	high-level language
compiler	input/output (I/O)
computer	program
decomposition	software
FORTRAN 77	stepwise refinement
Fortran 90	syntax
hardware	top-down design

References

For further reading on the 10 top engineering and scientific achievements of the last 25 years, we recommend the following references. Many of these references are from the National Academy of Engineering brochure entitled "10 Outstanding Achievements 1964–1989," published in 1989.

MOON LANDING

Hallion, Richard P., and Tom D. Crouch, eds. "Apollo: Ten Years Since Tranquility Base." Washington, D.C.: Smithsonian Institution Press, 1979.
Stix, Gary, ed. "Moon Lander." Spectrum, Vol. 25, No. 11, 1988, pp. 76–82.

APPLICATION SATELLITES

Canby, Thomas Y. "Satellites That Serve Us." National Geographic, September 1983, pp. 281–334.
Heckman, Joanne. "Read, Set, GOES: Weather Eyes for the 21st Century." Space World, July 1987, pp. 23–26.

MICROPROCESSORS

Garetz, Mark. "Evolution of the Microprocessor: An Informal History." BYTE, September 1985, pp. 209-215.

Toong, Hou-Min D. "Microprocessors." Scientific American, Vol. 237, No. 3, 1977, pp. 146-151.

COMPUTER-AIDED DESIGN AND MANUFACTURING

Loeffelholz, Suzanne. "CAD/CAM Comes of Age." Financial World, October 18, 1988, pp. 38-40.

Mitchell, Larry D. "Computer-Aided Design and Manufacturing." McGraw-Hill Encyclopedia of Science & Technology. New York: McGraw-Hill, 1987.

CAT SCAN

Andreasen, Nancy C. "Brain Imaging: Applications in Psychiatry." Science, March 18, 1988, pp. 1381-1388.

Sochurek, Howard. "Medicine's New Vision." National Geographic, January 1987, pp. 2-40.

ADVANCED COMPOSITE MATERIALS

Chou, Tsu-Wei, Roy L. McCullough, and R. Byron Pipes, "Composites." Scientific American, October 1986, pp. 192-203.

Steinberg, Morris A. "Materials for Aerospace." Scientific American, October 1986, pp. 67-72.

JUMBO JETS

Ingells, Douglas J. "747: Story of the Boeing Super Jet." Fallbrook, Calif: Aero Publishers, 1970.

Stewart, Stanley. "Flying the Big Jets." New York: Arco Publishing, 1985.

LASERS

"Lasers Then and Now" (special issue). Physics Today, October 1988.

Townes, Charles H. "Harnessing Light." Science 84, November 1984, pp. 153-155.

FIBER-OPTIC COMMUNICATIONS

Bell, Trudy, ed., "Fiber Optics." Spectrum, Vol. 25, No. 11, 1988, pp. 97-102.

Lucky, Robert W. "Message by Light Wave." Science 85, November 1985, pp. 112-113.

GENETICALLY ENGINEERED PRODUCTS

Eskow, Dennis. "Here Come the Assembly-Line Genes." Popular Mechanics, March 1983, pp. 92-96.

Weaver, Robert F. "Beyond Supermouse: Changing Life's Genetic Blueprint." National Geographic, December 1984, pp. 818-847.

2 Arithmetic Computations

Stride Estimation The motion of a simple pendulum can be used to model the motion of a leg in a natural stride. Using this model, it can be shown that the time that it takes for a stride is proportional to the length of the leg. Thus, the longer the leg, the longer the stride. If we know the length of a stride and the time it takes to make a stride, we can compute the walking speed, which is the number of feet per second traveled while walking. As you might have guessed, people with longer legs have a faster walking speed and go farther per second. These results do not apply to running because the leg does not swing freely in a running gait, and thus it cannot be modeled by a simple pendulum. With an accurate model of a leg, new prosthetic legs can be designed using advanced composite materials which are lightweight and strong.

INTRODUCTION

Arithmetic operations (adding, subtracting, multiplying, and dividing) are the most fundamental operations performed by computers. Engineers and scientists also need other routine operations, such as raising a number to a power, taking the logarithm of a number, or computing the sine of an angle. This chapter discusses methods of storing data with FORTRAN 77 and develops the statements for performing arithmetic calculations with that data. We also introduce statements for simple data input and output. With this group of statements, we can write complete FORTRAN 77 programs.

2-1 CONSTANTS AND VARIABLES

Numbers are introduced into a computer program either directly with the use of *constants* or indirectly with the use of *variables.* Constants are numbers used in FORTRAN 77 statements, such as -7, 3.141593, and 32.0. Constants may contain plus or minus signs and decimal points, but they may not contain commas. Thus, 3147.6 is a valid FORTRAN constant, but 3,147.6 is not. Constants are stored in memory locations but can be accessed only by using the constant value itself.

A variable represents a memory location that is assigned a name. The memory location contains a value; we reference that value with the name assigned to that memory location. We can visualize variables, their names, and their values as shown:

AMOUNT	**36.84**		VOLUME	**183.0**
RATE	**0.065**		TOTAL	**486.5**
TEMP	**17.5**		INFO	**72**

Each memory location to be used in a program is given a name and may be assigned a value using a FORTRAN statement. In the preceding example the memory location named RATE has been assigned the value 0.065.

Variable Names

Each variable must have a different name, which you provide in your program. The names may contain one to six characters consisting of both alphabetic characters and digits; however, the first character of a name must be alphabetic. FORTRAN does not distinguish between uppercase and lowercase characters. In all examples in this module, FORTRAN variables and statements appear in uppercase characters. The following are examples of both valid and invalid variable names:

Variable Name	Valid or Invalid
DISTANCE	invalid — too long
TIME	valid
PI	valid
$	invalid — illegal character ($)
TAX-RT	invalid — illegal character (-)
B1	valid
2X	invalid — first character must be alphabetic

Data Types

FORTRAN 77 allows you to use six different types of values, as illustrated in the following list:

Data Type	Examples
Integers	32, −7
Real values	−15.45, 0.004
Double-precision values	3.1415926536, 1.000000006
Complex values	$1 - 2i, 5i$ (where $i = \sqrt{-1}$)
Character values	'velocity', 'Report 3'
Logical values	.TRUE., .FALSE.

The first four types of values represent numerical values. This chapter focuses on integers and real values; double-precision and complex values are discussed in Chapter 8. Logical constants and variables are discussed in Chapter 3; character constants and variables are discussed in Chapter 8.

Integer values are those with no fractional portion and no decimal point, such as 16, −7, 186, and 0. On the other hand, *real values* contain a decimal point and may or may not have digits past the decimal point, such as 13.86, 13., 0.0076, −14.1, 36.0, and −3.1; real values are also called floating point values.

A memory location can contain only one type of value. The type of value stored in a variable is specified by two methods: *implicit typing* or *explicit typing.* With implicit typing, the first letter of a variable name determines the value type that can be stored in it. Variable names beginning with the letters *I, J, K, L, M,* or *N* are used to store integers. Variable names beginning with one of the other letters, *A - H* and *O - Z,* are used to store real values. Thus, with implicit typing, AMOUNT represents a real value and MONEY represents an integer value. An easy way to remember which letters are used for integers is to observe that the range of letters is *I-N,* the first two letters of the word *integer.*

With explicit typing, FORTRAN statements are used to specify the variable types. For example, the statements

```
INTEGER WIDTH
REAL ITEM, LENGTH
```

specify that WIDTH is a variable containing an integer value and that ITEM and LENGTH are variables containing real values. These *specification statements* or *type statements* have the following general forms:

INTEGER *variable list*

REAL *variable list*

The variable list contains variable names separated by commas. These statements are *nonexecutable* because they are not translated into machine language. Instead, they are used by the compiler to assign memory locations and to specify the types of values to be stored in the locations.

You will find it helpful to select a variable name that is descriptive of the value being stored. For example, if a value represents a tax rate, name it RATE or TAX. If the implicit typing of the variable name does not match the type of value to be stored in it, then use a REAL or INTEGER statement at the beginning of your program to specify the desired type of value. It is good programming style to list all variables in specification statements, including those correctly typed by the implicit rules. In the example programs used in this module, all variables are included in specification statements.

The PARAMETER statement is a specification statement used to assign names to constants, with the following general form:

> PARAMETER *(name1 = expression, name2 = expression, . . .)*

The expression after the equal sign typically is a constant, although it can be an expression consisting of other constants and operations such as those discussed later in this chapter. An example of a PARAMETER statement is

PARAMETER (PI=3.141593)

Type statements should precede the PARAMETER statement in order to assign the proper type to the constant name.

Scientific Notation

When a real number is very large or very small, decimal notation does not work satisfactorily. For example, a value that is used frequently in chemistry is Avogadro's constant, whose value to four significant places is 602,300,000,000,000,000,000,000. Obviously, we need a more manageable notation for very large values like Avogadro's constant or for very small values like 0.000000000042. *Scientific notation,* commonly used in science, expresses a value as a number between 1 and 10 multiplied by a power of 10. In scientific notation Avogadro's constant becomes 6.023×10^{23}. Elements of this form are commonly referred to as the mantissa (6.023) and the exponent (23). The FORTRAN form of scientific notation, called *exponential notation,* expresses a value as a number between 0.1 and 1 multiplied by an appropriate power of 10. Exponential notation uses the letter E to separate the mantissa and the exponent. In exponential form Avogadro's constant becomes 0.6023E24. The following are other examples of decimal values in scientific and exponential notation:

Decimal	Scientific	Exponential
3,876,000,000	3.876×10^{9}	0.3876E10
0.0000010053	1.0053×10^{-6}	0.10053E$-$05
$-8,030,000$	-8.03×10^{6}	$-0.803E07$
-0.000157	-1.57×10^{-4}	$-0.157E-03$

Although FORTRAN uses an exponential form with a mantissa between 0.1

Table 2-1   Integer Representations in Typical Computers

Computer	Number of Binary Digits (or Bits) per Data Value	Number of Integers*
Texas Instruments 32020 Microprocessor	16	65,536
Motorola 68020 Microprocessor	32	4.3×10^9
IBM PC	32	4.3×10^9
Macintosh	32	4.3×10^9
VAX 11/780	32	4.3×10^9
Sun Sparc Station 10	32	4.3×10^9
Cray Y MP C90	48	2.8×10^{14}
Cray-2	64	1.8×10^{19}

* Generally, half of the integers will represent negative values, so the maximum integer is usually equal to the number of integers divided by 2.

and 1.0, it will accept mantissas outside that range; for instance, the constant 0.16E03 would also be valid in the forms 1.6E02 or 0.016E04.

Magnitude Limitations

The magnitude and precision of values that can be stored in a computer are limited. All limitations on values depend on the specific computer. For instance, π is an irrational number and cannot be written with a finite number of decimal positions; in a computer with seven digits of accuracy, π could be stored as 3.141593. In addition to limits on the number of significant positions in the mantissa, there are also limits on the size of the exponent.

Table 2-1 compares the approximate range of integers that can be stored in several computers. The range of values is determined by the design of the central processing unit (CPU). Check a reference manual to find the ranges of real and integer values for the computer you will be using.

Try It

Try this self-test to check your memory of some key points from Section 2-1. If you have any problems with the exercises, you should reread this section. The solutions are given at the end of this module.

Problems 1–10 contain both valid and invalid variable names. Explain why the invalid names are unacceptable.

1. SQ.YD
2. MICRON
3. WEIGHT
4. DEGREES
5. NET__WT
6. SIDE-1
7. F(T)
8. 3J6
9. TOTAL
10. FIVE%

In problems 11–16, tell whether or not the pair of real constants represents the same number. If not, explain.

11. 15.7; 0.157E–2
12. –1.7; 1.7E–01
13. 10; 1000.0E–02
14. 0.005; 0.00005E–02
15. 0.899; 89.9E02
16. –0.044; 4.4E–02

2-2 ARITHMETIC OPERATIONS

Computations in FORTRAN may be specified with the *assignment statement,* which has the general form

> *variable name = expression*

The simplest form of an expression is a constant. Hence, if the value of π is needed frequently in a program, we might choose to define a variable PI with the value 3.141593. We could then refer to the variable PI each time we needed the constant. An assignment statement that assigns a value to PI is

PI = 3.141593

The name of the variable receiving a new value must always be on the left side of the equal sign. In FORTRAN the equal sign can be read as "is assigned the value of." Thus, this statement could be read "PI is assigned the value 3.141593." The term *initialized* is often used to refer to the first value assigned to a variable in a program; this statement could also be read "PI is initialized to the value 3.141593."

Simple Arithmetic

Often we want to calculate a new value using arithmetic operations with other variables and constants. For instance, assume that the variable RADIUS has been assigned a value and we want to calculate the area of a circle having that radius. To do so, we must square the radius and then multiply by the value of π. Table 2-2 shows the FORTRAN expressions for the basic arithmetic operations. Note that an asterisk (instead of $\times$) represents multiplication; this avoids confusion, because A x B (commonly used in algebra to indicate the product of A and B) represents a variable name in FORTRAN. Division and exponentiation also have different symbols that allow us to write these arithmetic operations on a single line.

Evaluating an Arithmetic Expression

Because several operations can be combined in one *arithmetic expression,* it is important to determine the priorities of the operations — that is, the order in which the operations are performed. For instance, consider the following assignment statement that calculates the area of a circle:

AREA = PI*RADIUS**2

If the exponentiation is performed first, we compute PI $\cdot$ (RADIUS)^2; if multiplication is performed first, we compute (PI $\cdot$ RADIUS)^2. Note that the two computations yield different results. The order of priorities for computations in FORTRAN is given in Table 2-3 and follows the standard algebraic priorities.

When executing the previous FORTRAN statement, the RADIUS is first squared; then the result is multiplied by PI — correctly determining the area of the circle. Remember that we assume that both PI and RADIUS have been initialized.

Table 2-2   Arithmetic Operations in Algebraic Form and in FORTRAN

Operation	Algebraic Form	FORTRAN
Addition	A + B	A + B
Subtraction	A − B	A − B
Multiplication	A × B	A*B
Division	$\dfrac{A}{B}$	A/B
Exponentiation	A^3	A**3

If a minus sign precedes the first variable name in an expression, it is computed on the same priority level as subtraction. For example, −A**2 is computed as if it were −(A**2), −A*B is computed as if it were −(A*B), and −A + B is computed as if it were (−A) + B.

When two operations are on the same priority level, as in addition and subtraction, all operations except exponentiation are performed from left to right. Thus, B − C + D is evaluated as (B − C) + D. If two exponentiations occur sequentially in FORTRAN, as in A**B**C, they are evaluated right to left, as in A**(B**C). Thus, 2**3**2 is 2^9, or 512, as opposed to (2**3)**2, which is 8^2, or 64.

A more complicated example is represented by the following equation for one of the real roots of a quadratic equation:

$$X1 = \frac{-B + \sqrt{B^2 - 4AC}}{2A}$$

A, B, and C are coefficients of the quadratic equation ($AX^2 + BX + C = 0$). Because computers cannot divide by zero, we will assume for now that A is not equal to zero. The value of X1 can be computed in FORTRAN with the following statement, assuming that the variables A, B, and C have been initialized:

```
X1 = (–B + (B**2 – 4.0*A*C)**0.5)/(2.0*A)
```

To check the order of operations in a long expression, we should start with the operations inside parentheses; that is, find the operation done first, then second, and so on. The following diagram outlines this procedure, using braces to show the steps of operations. Beneath each brace is the value calculated in that step:

$$\begin{array}{l} X1 = \underbrace{(-B}_{-B} + \underbrace{\overbrace{(B**2}^{B^2} - \overbrace{4.0*A*C)}^{4AC}**0.5)}_{B^2 - 4AC}/\underbrace{(2.0*A)}_{2A} \\ \underbrace{\qquad\qquad -B + \sqrt{B^2 - 4AC}\qquad\qquad} \\ \dfrac{-B + \sqrt{B^2 - 4AC}}{2A} \end{array}$$

Table 2-3 Priorities of Arithmetic Operations

Priority	Operation
1	Parentheses
2	Exponentiation
3	Multiplication and division
4	Addition and subtraction

As shown in the final brace, the desired value is computed by this expression.

Parentheses placement is important. If the outside set of parentheses in the numerator in the previous FORTRAN statement was omitted, the assignment statement would become

$$X1 = \underbrace{-B}_{-B} + \overbrace{\underbrace{\underbrace{\cfrac{\underbrace{\overbrace{(B**2 - 4.0*A*C)}^{B^2 - 4AC}}_{\sqrt{B^2 - 4AC}}**0.5}{\overbrace{(2.0*A)}^{2A}}}_{\cfrac{\sqrt{B^2 - 4AC}}{2A}}}_{-B + \cfrac{\sqrt{B^2 - 4AC}}{2A}}}$$

As you can see, omission of the outside set of parentheses causes the wrong value to be calculated as a root of the original quadratic equation. Omitting necessary parentheses results in incorrect calculations. Using extra parentheses to emphasize the order of calculations is permissible, even though they may not be needed. We can insert extra parentheses in a statement to make the statement more readable.

You also may want to break a long statement into several smaller statements. The expression $B^2 - 4AC$ in the quadratic equation is called the discriminant. Both roots of the solution could be calculated with the following statements after initialization of A, B, and C:

```
DISCR = B**2 - 4.0*A*C
X1 = (-B + DISCR**0.5)/(2.0*A)
X2 = (-B - DISCR**0.5)/(2.0*A)
```

In the preceding statements we assume that the discriminant, DISCR, is positive, enabling us to obtain X1 and X2, the two real roots to the equation. If the discriminant were negative, an execution error would occur when we attempted to take the square root of the negative value. If the value of A were zero, we would get an execution error for attempting to divide by zero. Later chapters present techniques for handling these situations.

Truncation and Mixed-Mode Operations

When an arithmetic operation is performed using two real numbers, its *intermediate result* is a real value. For example, we can calculate the circumference of a circle using either of the following two statements:

```
CIRCUM = PI*DIAMTR
CIRCUM = 3.141593*DIAMTR
```

In both statements, we have multiplied two real values, giving a real result, which is then stored in the real variable CIRCUM.

Similarly, arithmetic operations between two integers yield an integer. For instance, if I and J represent two integers and if I is less than or equal to J, then we can calculate the number of integers in the interval [I,J] with the following statement:

```
INTERV = J - I + 1
```

Thus, if I = 6 and J = 11, INTERV will be assigned the value 6, the number of integers in the set {6, 7, 8, 9, 10, 11}.

Now consider the statement

```
LENGTH = SIDE*3.5
```

Assume that SIDE represents a real value and that LENGTH represents an integer value. We know that the multiplication between the real value SIDE and the real constant 3.5 yields a real result. In this case, however, the real result is stored in an integer variable. When the computer stores a real number in an integer variable, it ignores the fractional portion and stores only the whole number portion of the real number; this type of rounding is called *truncation.*

Computations with integers can also give unexpected results. Consider the following statement, which computes the average, or mean, of two integers, N1 and N2:

```
MEAN = (N1 + N2)/2
```

If we assume that all the variables in the statement are integers, the result of the expression will be an integer. Thus, if N1 = 2 and N2 = 4, the mean value is the expected value, 3. But if N1 = 2 and N2 = 3, the result of the division of 5 by 2 will be 2 instead of 2.5 because the division involved two integers; hence, the intermediate result must be an integer. At first glance it might seem that we could solve this problem if we called the average by a real variable named AVE (instead of MEAN) and used this statement:

```
AVE = (N1 + N2)/2
```

Unfortunately, this cannot correct our answer. The result of integer arithmetic is still an integer; all we have done is to move the integer result into a real variable. Thus, if N1 = 2 and N2 = 3, then (N1 + N2)/2 = 2 and AVE = 2.0, not 2.5. One way to correct this problem is to declare N1 and N2 to be real values and use the following statement to calculate the average:

```
AVE = (N1 + N2)/2.0
```

Note the difference between *rounding* and truncation. With rounding the result is the integer closest in value to the real number. Truncation, however, causes any decimal portion to be dropped. If we divide the integer 15 by the integer 8, the truncated result is 1, and the rounded result is 2.

The effects of truncation can also be seen in the following statement, which appears to calculate the square root of NUM:

```
ROOT = NUM**(1/2)
```

However, since 1/2 is truncated to 0, we are really raising NUM to the zero power; ROOT will always contain the value 1.0, no matter what value is in NUM.

We have seen that an operation involving only real values yields a real result, and an operation involving only integer values yields an integer result. FORTRAN also accepts a *mixed-mode operation,* which is an operation involving an integer value and a real value. The intermediate result is a real value. The final result depends on the type of variable used to store the result of the mixed-mode operation. Consider the following arithmetic statement for computing the perimeter of a square whose sides are real values:

```
PERIM = 4*SIDE
```

The preceding multiplication is a mixed-mode operation between the integer constant 4 and the real variable SIDE. The intermediate result is real and is correctly stored in the real variable PERIM.

Using mixed mode, we can now correctly calculate the square root of the integer NUM, using this statement:

```
ROOT = NUM**0.5
```

The mixed-mode exponentiation yields a real result, which is stored in ROOT.

To compute the area of a square with real sides, we could use either of the following statements:

```
AREA = SIDE**2
AREA = SIDE**2.0
```

The result in both cases is real, but the first form is preferable because exponentiation to an integer power is generally performed internally in the computer with a series of multiplications such as SIDE times SIDE. If an exponent is real, however, the operation is performed by the arithmetic logic unit using logarithms; SIDE**2.0 is actually computed as antilog $(2.0 \times \log(\text{SIDE}))$. Logarithms can introduce small errors into the calculations; although 5.0**2 is always 25.0, 5.0**2.0 is often computed as 24.99999. Also, note that (-2.0)**2 is a valid operation, but (-2.0)**2.0 is an invalid operation — the logarithm of a negative value does not exist, and an execution error occurs. As a general guide when raising numbers to an integer power, use an integer exponent, even though the base number is real.

Assume that we want to calculate the volume of a sphere with radius R, where R represents a real value. The volume is computed by multiplying 4/3 times π, times the radius cubed. The following mixed-mode statement at first appears correct:

$$\text{VOLUM} = (4/3)*3.141593*R**3$$

The expression contains integer and real values, so the result will be a real value. However, the division of 4 by 3 yields the intermediate value of 1, not 1.333333; therefore, the final answer will be incorrect.

Because mixed-mode operations can sometimes give unexpected results, you should try to avoid writing arithmetic expressions that include them. A mixed-mode expression is desirable in an exponentiation operation in which a real value is raised to an integer power.

Underflow and Overflow

In a previous section we discussed magnitude limitations for the values stored in variables. Because the maximum and minimum values that can be stored in a variable depend on the computer system itself, a computation may yield a result that can be stored in one computer system but is too large to be stored in another. For example, suppose we execute the following assignment statements:

$$X = 0.25E20$$
$$Y = 0.10E30$$

These values are both valid for a computer with an exponent range of -38 through 38. Suppose we now execute the following statement:

$$Z = X*Y$$

The numerical result of this multiplication is .025E50. Clearly this result is too large to store in a computer with a maximum exponent of 38. The error is not a syntax error; the statements themselves are valid FORTRAN 77 statements. The error is an execution error because it occurred during execution of the program, and it is called an exponent *overflow* error because the exponent of the result of an arithmetic operation was too large to store in the computer's memory.

Exponent *underflow* is a similar error caused by the exponent of the result of an arithmetic operation being too small to store in the computer's memory. Using the same computer as we did for the previous example, the following statements would generate an exponent underflow error because the value that is computed for C has an exponent smaller than -38:

$$A = 0.25E-20$$
$$B = 0.10E+20$$
$$C = A/B$$

If you get exponent underflow or overflow errors when you run your programs, you need to examine the magnitude of the values you are using. If you really need values whose exponents exceed the limits of your computer, the only solution is to switch to a computer that can handle a wider range of exponents. Most of the time, exponent underflow and overflow errors are caused by other errors in a program. For example, if variables are initialized to an incorrect value or the wrong arithmetic operation is specified, exponent underflow or overflow can occur, but the source of the problem is elsewhere.

Try It

Try this self-test to check your memory of some key points from Section 2-2. If you have any problems with the exercises, you should reread this section. The solutions are given at the end of this module.

1. What value is stored in Y after the following statements are executed?

```
REAL X, A, Y
X = 5.0
A = 2.0
Y = X + 3.0/A*5.0
```

2. What value is stored in X after the following statements are executed?

```
REAL X, X1, X2
X1 = 4.0
X2 = 10.0
X = 1.0/(1.0/X1) + (1.0/X2)
```

3. What value is stored in RESULT after the following statements are executed?

```
INTEGER X, Y, B, RESULT
X = 5
Y = -21
B = 16
RESULT = X + Y + B/5
```

2-3 INTRINSIC FUNCTIONS

Algorithms commonly require simple operations, such as computing the square root of a value, computing the absolute value of a number, or computing the sine of an angle. Because these operations occur so frequently, built-in computer functions called *intrinsic functions* are available to handle these routine computations. Instead of using the arithmetic expression X**0.5 to compute a square root, we can use the intrinsic function SQRT(X). Similarly, we can refer to the absolute value of B by ABS(B). A list of some commonly used intrinsic functions appears in Table 2-4. Appendix A contains a complete list of FORTRAN 77 intrinsic functions and a brief description of each.

The name of a function is followed by the input to the function, called the *argument* of the function, which is enclosed in parentheses. This argument can be a constant, variable, or expression. For example, suppose we want to compute the cosine of the variable ANGLE and store the result in another variable COSINE. From Table 2-4 we see that the cosine function assumes that its argument is in radians. If the value in ANGLE is in degrees, we must change the degrees to radians (1 degree = $\pi/180$ radians) and then compute the cosine. The following statement performs the degree-to-radian conver-

Table 2-4   Common Intrinsic Functions

Function Name and Argument	Function Value	Comments
SQRT(X)	$\sqrt{X}$	Square root of X
ABS(X)	$|X|$	Absolute value of X
SIN(X)	Sine of angle X	X must be in radians
COS(X)	Cosine of angle X	X must be in radians
TAN(X)	Tangent of angle X	X must be in radians
EXP(X)	e^X	e raised to the X power
LOG(X)	$\log_e X$	Natural log of X
LOG10(X)	$\log_{10} X$	Common log of X
INT(X)	Integer part of X	Converts a real value to an integer value
REAL(I)	Real value of I	Converts an integer value to a real value
MOD(I,J)	Integer remainder of I/J	Remainder or modulo function

sion within the function argument. Note that the inside set of parentheses is not required but emphasizes the conversion factor.

$$\text{COSINE} = \text{COS(ANGLE*(3.141593/180.0))}$$

The REAL and INT functions may be used to avoid undesirable mixed-mode arithmetic expressions by explicitly converting variable types. For example, if we are computing the average of a group of real values, we need to divide the real sum of the values by the number of values. We can convert the integer number of values into a real value for the division using the REAL function, as shown in the following statement:

$$\text{AVERG} = \text{SUM/REAL(N)}$$

If the values were integers, the sum would also be an integer. We probably would still want the average to be represented by a real value, which could be specified by

$$\text{AVERG} = \text{REAL(SUM)/REAL(N)}$$

Many intrinsic functions are *generic functions,* which means that the value returned is the same type as the input argument. The absolute value function ABS is a generic function. If X is an integer, then ABS(X) is also an integer; if X is a real value, then ABS(X) is also a real value. Some functions specify the type of input and output required. IABS is a function that requires an integer input and returns an integer absolute value from the function. If K is an integer, then ABS(K) and IABS(K) return the same value. Appendix A contains all forms of the intrinsic functions and identifies all generic functions.

Try It

Try this self-test to check your memory of some key points from Sections 2-2 and 2-3. If you have any problems with the exercises, you should reread these sections. The solutions are included at the end of this module.

In problems 1–4 convert the equations into FORTRAN assignment statements. Assume all variables represent real values.

1. Magnitude:

$$M = \sqrt{x^2 + y^2}$$

2. Velocity addition formula:

$$u = \frac{u + v}{1 + \dfrac{u \cdot v}{c^2}}$$

3. Damped harmonic motion:

$$y = y_0 \cdot e^{-at} \cdot \cos(2\pi f \cdot t)$$

4. Temperature conversion to degrees K:

$$T = \left(\frac{5}{9}(T_f - 32)\right) + 273.15$$

In problems 5–10, convert the FORTRAN statements into algebraic form.

5. Potential energy:

```
PE = -G*ME*M/R
```

6. Electric flux:

```
DF = E*DA*COS(THETA)
```

7. Average velocity:

```
AV = (X2 - X1)/(T2 - T1)
```

8. Centripetal acceleration:

```
PI = 3.141593
CA = 4.0*PI**2*R/(T**2)
```

9. Distance of an accelerating body:

```
DIST = V*TIME + ACC*TIME**2/2.0
```

10. Atmospheric pressure adjusted for elevation:

```
P = P0*EXP(-M*G*X/R*TK)
```

2-4 SIMPLE INPUT AND OUTPUT

Before we discuss input and output statements, we outline the proper form for entering FORTRAN statements in a program. We then present several types of I/O statements. A computer can accept input from different sources, but in this chapter we assume that the input is from the keyboard. Similarly, we can direct output to different devices, but in this chapter we assume that the output is directed to a terminal screen or a printer. Input and output using a data file are discussed in Chapter 4.

FORTRAN Statement Format

We can create FORTRAN programs using an editor or a word processor to enter the text. Each statement is entered on a new line. We refer to specific positions within the line by column number: the first position in a line is column 1, the second position in a line is column 2, and so on.

Columns 1 through 5 are reserved for *statement numbers* (also called labels), which must be nonzero positive integers. Most statements do not require a statement number, but as you will see in later chapters, some statements need numbers so they can be referenced by other statements.

Column 6 is used to indicate that a statement has been continued from the previous line. Any nonblank character except a zero can be used in column 6 to indicate continuation. A FORTRAN statement may have several *continuation lines* if it is too long to fit on one line.

A FORTRAN statement starts in column 7 and can extend to column 72. In general, blanks can be inserted anywhere in the statement for readability. All information beyond column 72 is ignored. These rules for the general form of a FORTRAN statement are summarized in the following diagram:

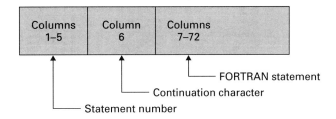

The only exception to the rules for spacing in FORTRAN statements applies to *comment lines.* Comment lines are used for entering general comments about the program, and they are printed in a listing of the program; comment lines are not converted into machine language or used during execution of the program. An asterisk or the letter C in column 1 indicates that a line is a comment line. It is good programming practice to include several comment lines near the beginning of the program to describe its purpose. In fact, you might want to include comments containing the information in step 1 (problem statement) and step 2 (input/output description) of the problem-solving process. Blank lines can also be included anywhere in the program.

Some terminal editors require you to number each line. These line numbers are used in editing but cannot be referenced by a FORTRAN statement. If you are using one of these editors, all statements will have a line number, and statements referenced in your program will also need a FORTRAN label.

List-Directed Output

FORTRAN has two types of statements that allow us to perform I/O operations. *List-directed* input/output statements are easy to use but give us little control over the exact spacing used in the input and output lines. *Formatted* input/output, although more involved, allows us to control the input and output forms with greater detail. This section presents list-directed input/output and simple formatted output. For the example programs in this section, we will use list-directed input statements and formatted output statements. This technique allows us to be flexible about the form we use for information we read and, at the same time, allows us to be specific about the form used to display the information computed by our program.

If we wish to print the value stored in a variable, we must tell the computer the variable's name. The computer can then access the memory location and print its contents. The general form of the list-directed PRINT statement is

> PRINT*, expression list

The expressions in the list must be separated by commas. The corresponding values are printed in the order in which they are listed in the PRINT statement. The output from each PRINT statement begins on a new line.

In our examples computer output is shown inside a rounded box, as illustrated in Example 2-1.

EXAMPLE 2-1 ## Weight, Vol, and Densty

Print the stored values of the variables WEIGHT, VOL and DENSTY.

SOLUTION

Computer Memory

WEIGHT	**35000**
VOL	**3.15**
DENSTY	**0.0000156**

FORTRAN Statement

```
        PRINT*, WEIGHT, VOL, DENSTY
```

Computer Output

```
35000   3.15000   0.156000E-04
```

. . . .

The number of decimal positions printed for real values and the spacing between items vary depending on the compiler used. If a value to be printed is very large or very small, many compilers automatically print the value in exponential notation instead of in decimal form.

Descriptive information (sometimes called literal information or *literals*) may also be included in the expression list by enclosing the information in single quotation marks or apostrophes. An apostrophe in a literal is represented by two quote marks, as in 'USER''S PROGRAM'. The descriptive information is then printed on the output line along with the values of any variables.

EXAMPLE 2-2 ## Literal and Variable Information

Print the variable RATE with a literal that identifies the value as representing a flow rate in gallons per second.

SOLUTION

Computer Memory   RATE `0.065`

FORTRAN Statement

```
      PRINT*, 'FLOW RATE IS', RATE, 'GALLONS PER SECOND'
```

Computer Output

```
FLOW RATE IS 0.0650000 GALLONS PER SECOND
```

. . . .

List-Directed Input

We frequently want to read information with our programs. The general form of a list-directed READ statement is

```
READ*, variable list
```

The variable names in the list must be separated by commas. The variables receive new values in the order in which they are listed in the READ statement. These values should agree in type (integer or real) with the variables in the list. The system will wait for you to enter the appropriate data values when the READ statement is executed. If more than one value is being read by the statement, the data values can be separated by commas or blanks.

A READ statement will read as many lines as needed to determine values for the variables in its list. Therefore, if a READ statement has four variables on it, it will wait until you have entered four values; the values could be on one line (separated by commas or blanks) or on several lines. Also, each READ statement begins reading from a new line. Thus, if you have executed a READ statement with four variables in its list and you entered five values on one line, the last value will not be read; the next READ statement will assume that a new line is used for entering information.

EXAMPLE 2-3   ## Capacitor Charge

Read the beginning and ending charges for a capacitor.

SOLUTION 1

This solution uses one READ statement; thus, both values can be entered on the same line.

FORTRAN Statement

```
        READ*, BEGIN, ENDING
```

Data Line

```
        186.93, 386.21
```

Computer Memory

BEGIN `186.93`

ENDING `386.21`

The data values in this example could also have been entered on separate lines since the READ statement will read as many lines as needed to find values for the variables in its list.

SOLUTION 2

This solution uses two READ statements; thus, the values must be entered on different lines.

FORTRAN Statements

```
READ*, BEGIN
READ*, ENDING
```

Data Lines

```
186.93
386.21
```

Computer Memory

```
BEGIN    186.93
ENDING   386.21
```

. . . .

Formatted Output

To specify the form in which data values are printed and where on the output line they are printed requires formatted statements. The general form of a formatted PRINT statement is

> PRINT *k, expression list*

The expression list designates the memory locations whose contents will be printed or arithmetic expressions whose values will be printed. The list of expressions also determines the order in which the values will be printed. The reference *k* refers to a FORMAT statement which will specify the spacing to be used in printing the information. A sample PRINT and FORMAT combination is

```
PRINT 5, TIME, DISTNC
5 FORMAT (1X,F5.1,2X,F7.2)
```

Recall that statement labels are entered in columns 1 through 5.
The general form of the FORMAT statement is

> *k* FORMAT *(specification list)*

The specification list tells the computer both the vertical and horizontal spacing to be used when printing the output information. The vertical spacing options include printing on the top of a new page (if the output is being printed on paper), the next line (single spacing), double spacing, and no spacing. Horizontal spacing includes indicating how many digits will be used for each value, how many blanks will be between numbers, and how many values are to be printed per line.

To understand the specifications used to describe the vertical and horizontal spacing, we must first examine the output from a printer or terminal, the most common output devices. Other forms of output have similar characteristics.

The line printer prints on continuous computer paper that is on a perforated roll so it is easy to separate the pages; a laser printer uses individual sheets of paper. Typically, 55 to 75 lines of information can be printed per page. The number of characters printed per line depends on the font size, but a typical line contains 60 to 70 characters. The PRINT/FORMAT combination specifically describes where each line is to be printed on the page (vertical spacing) and which positions in the line will contain data (horizontal spacing).

The computer uses the specification list to construct each output line internally in memory before actually printing the line. This internal memory region is called a *buffer.* The buffer is automatically filled with blanks before it is used to construct a line of output. The first character of the buffer is called the *carriage control character;* it determines the vertical spacing for the line. The remaining characters represent the line to be printed, as shown in the following diagram:

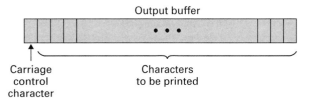

The following list shows some of the valid carriage control characters and the vertical spacing they generate. When needed for clarity in either FORMAT statements or buffer contents, a blank is indicated by the character b placed one-half space below the regular line.

Carriage Control Character	Vertical Spacing
1	New page
blank	Single spacing
0	Double spacing
+	No vertical spacing

Double spacing causes one line to be skipped before the current line of output is printed. When a plus sign is in the carriage control, no spacing occurs and the next line of information will print over the last line printed. On most computers an invalid carriage control character causes single spacing.

A terminal screen does not have the same capabilities as a printer for spacing; therefore, a 1 in the carriage control usually becomes an invalid control character and causes single spacing. If the terminal I/O does not use carriage control, then the entire contents of the buffer, including the carriage control character, may appear on the terminal screen.

We now examine four FORMAT specifications that describe how to fill the output buffer. Commas are used to separate specifications in the FORMAT statement. Additional FORMAT specifications are included in Section 2-8.

Literal Specification The literal specification allows us to put characters directly into the buffer. The characters must be enclosed in single quotation marks or apostrophes. These characters can represent the carriage control character or the characters in a literal. The following examples illustrate use of the literal specification in FORMAT statements.

EXAMPLE 2-4 ## Title Heading

Print the title heading TEST RESULTS on the top of a new page, *left-justified* (that is, no blanks to the left of the heading).

SOLUTION

FORTRAN Statements

```
      PRINT 4
    4 FORMAT ('1','TEST RESULTS')
```

Buffer Contents

```
1TEST RESULTS
```

Computer Output

```
        111
123456789012
```

```
TEST RESULTS
```

The buffer is filled according to the FORMAT. No variable names were listed on the PRINT statement; hence, no values were printed. The literal specifications cause the characters 1TEST RESULTS to be put in the buffer, beginning with the first position in the buffer. After filling the buffer, as instructed by the FORMAT, the carriage control is examined to determine vertical spacing. The character 1 in the carriage control position tells the computer to begin a new page. The rest of the buffer is then printed. Notice that the carriage control character is not printed. The row of small numbers above the computer output shows the specific column of the output line: the first T is in column 1, the second T is in column 4, and the third T is in column 11.

• •

EXAMPLE 2-5 ## Column Headings

Double space from the last line printed and print column headings 1988 kWh and 1989 kWh, with no blanks on the left side of the line and seven blanks between the two column headings.

CORRECT SOLUTION

FORTRAN Statements

```
      PRINT 3
    3 FORMAT ('0','1988 kWhbbbbbbb1989 kWh')
```

Buffer Contents

> 01988 kWhbbbbbbb1989 kWh

Computer Output

```
                  11111111112222
         12345678901234567890123
```

> 1988 kWh   1989 kWh

The line shown is printed after double spacing from the previous line of output.

INCORRECT SOLUTION

FORTRAN Statements

```
        PRINT 3
      3 FORMAT ('1988 kWhbbbbbbb1989 kWh')
```

Buffer Contents

> 1988 kWhbbbbbbb1989 kWh

Computer Output

```
                  1111111111222
         12345678901234567890012
```

> 988 kWh   1989 kWh

In this example we forgot to specify the carriage control. However, the computer does not forget: the first position of the buffer contains a 1, which indicates spacing to a new page. The rest of the buffer is then printed.

· ·

X Specification   The X specification will insert blanks into the buffer. Its general form is nX, where n represents the number of blanks to be inserted in the buffer. An example using both the X specification and the literal specification follows.

EXAMPLE 2-6   **Centered Heading**

Print the heading EXPERIMENT NO. 1 centered at the top of a new page.

SOLUTION

Assume that an output line contains 65 characters. To center the heading, we determine the number of characters in the heading (16), subtract that from 65 ($65 - 16 = 49$), and divide that by 2 to put one-half the blanks in front of the heading ($49/2 = 25$).

FORTRAN Statements

```
        PRINT 35
    35 FORMAT ('1',25X,'EXPERIMENT NO. 1')
```

Buffer Contents

```
1bbbbbbbbbbbbbbbbbbbbbbbbbEXPERIMENT NO. 1
```

Computer Output

```
        2222333333333344
        6789012345678901
```

```
EXPERIMENT NO. 1
```

. . . .

I Specification   The literal specification and the X specification allow us to specify carriage control and to print headings. They cannot, however, be used to print variable values. We now examine a specification that prints the contents of integer variables: the specification form is Iw, where w represents the number of positions (width) to be assigned in the buffer for printing the value of an integer variable. The value is always *right-justified* (no blanks to the right of the value) in those positions in the buffer. Extra positions on the left are filled with blanks. Thus, if the value 16 is printed with an I4 specification, the four positions contain two blanks followed by 16. If not enough positions are available to print the value, including a minus sign if the value is negative, the positions are filled with asterisks. Hence, if we print the value 132 or -14 with an I2 specification, the two positions are filled with asterisks. It is important to recognize that the asterisks do not necessarily indicate that there is an error in the value; instead, the asterisks may indicate that you need to assign a larger width in the corresponding output specification.

More than one variable name is often listed in the PRINT statement. When interpreting a PRINT/FORMAT combination, the compiler will match the first variable name to the first specification for printing values, and so on. Therefore, there should be the same number of specifications for printing values as there are variables on the PRINT statement list. (In Section 2-8 we explain what happens if the number of specifications does not match the number of variables.)

EXAMPLE 2-7

Integer Values

Print the values of the integer variables SUM, MEAN, and N on the same line, single spaced from the previous line.

SOLUTION

Computer Memory

SUM	12
MEAN	−14
N	−146

FORTRAN Statements

```
      PRINT 30, SUM, MEAN, N
30 FORMAT (1X,I3,2X,I2,2X,I4)
```

Buffer Contents

```
bb12bb**bb-146
```

Computer Output

```
           1111
  1234567890123
```

```
12   **   -146
```

The computer will print SUM with an I3 specification, MEAN with an I2 specification, and N with an I4 specification. The value of SUM is 12, so the three corresponding positions contain a blank followed by the number 12. The value of MEAN, -14, requires at least three positions, so the two specified positions are filled with asterisks. The value of N fills all four allotted positions. The carriage control character is a blank, thus the line of output is single spaced from the previous line.

. . . .

EXAMPLE 2-8

Literal and Variable Information

On separate lines print the values of MEAN and SUM, along with an indication of the name of each of the integer variables.

SOLUTION

Computer Memory

SUM `12`
MEAN `-14`

FORTRAN Statements

```
      PRINT 2, MEAN
2 FORMAT (1X,'MEAN = ',I4)
      PRINT 3, SUM
3 FORMAT (1X,'SUM = ',I4)
```

Buffer Contents

```
bMEANb=bb-14
```

```
bSUMb=bbb12
```

Computer Output

```
          11
  12345678901
```

```
MEAN =   -14
SUM =    12
```

. . . .

F Specification   The F specification is used to print real numbers in a decimal form (for example, 36.21) as opposed to an exponential form (for example, 0.3621E+02). The general form for an F specification is Fw.d, where w represents the total width (number of positions, including the decimal point) to be used, and d represents the number of those positions that will represent decimal positions to the right of the decimal point as shown below:

$$
\underbrace{XX.\overbrace{XXX}^{\text{decimal portion} = d}}_{\text{total width} = w}
$$

If the value to be printed has fewer than d decimal positions, zeros are inserted on the right side of the decimal point. Thus, if the value 21.6 is printed with an F6.3 specification, the output is 21.600. If the value to be printed has more than d decimal positions, only d decimal positions are printed, dropping the rest. Thus, if the value 21.86342 is printed with an F6.3 specification, the output is 21.863. Many compilers will round to the last decimal position printed; in these cases the value 18.98662 is printed as 18.987 if an F6.3 specification is used.

If the integer portion of a real value requires fewer positions than allotted in the F specification, the extra positions on the left side of the decimal point are filled with blanks. Thus, if the value 3.123 is printed with an F6.3 specification, the output is a blank followed by 3.123. If the integer portion of a real value, including the minus sign if the value is negative, requires more positions than allotted in the F specification, the entire field is filled with asterisks. Thus, if the value 312.6 is printed with an F6.3 specification, the output is ******.

If a value is between −1 and +1, positions must usually be allowed for both a leading zero to the left of the decimal point and a minus sign if the value is negative. Thus, the smallest F specification that could be used to print −0.127 is F6.3. If a smaller specification width were used, all the positions would be filled with asterisks.

EXAMPLE 2-9   Angle THETA

Print the value of an angle called THETA. Construct the Greek symbol for θ using a zero with a dash printed over it. The output should be in this form:

$$\theta = XX.XX$$

SOLUTION

Computer Memory

THETA   **3.184**

FORTRAN Statements

```
        PRINT 1
   1 FORMAT (1X,'0')
        PRINT 2, THETA
   2 FORMAT ('+','- =',F5.2)
```

Buffer Contents

b0

+-b=bb3.18

Computer Output

123456789

0 =   3.18

The first PRINT statement prints a zero in the first position of the output line after single spacing from the previous line. The second PRINT statement has a plus sign in the carriage control, which causes no vertical spacing. The dash character is therefore printed on top of the character zero, giving the Greek symbol theta. The value of the variable THETA is printed on the same line. (Remember that overprinting will work only if the output device uses carriage control; otherwise all output is single-spaced.)

· ·

E Specification   Real numbers may be printed in an exponential form with the E specification. This specification is used primarily for very small or very large values, or when you are uncertain of the magnitude of a number. If you use an F format that is too small for a value, the output field will be filled with asterisks. In contrast, a real number will always fit in an E specification field.

The general format for an E specification is Ew.d. The w again represents the total width or number of positions to be used in printing the value. The d represents the number of positions to the right of the decimal point, assuming that the value is in exponential form, with the decimal point to the left of the first nonzero digit. The framework for printing a real value in an exponential specification with three decimal places is

$$\underbrace{\text{S0.}\overbrace{\text{XXX}}^{\text{decimal portion} = d}\text{ESXX}}_{\text{total width} = w}$$

The symbol S indicates that positions must be reserved for both the sign of the value and the sign of the exponent in case they are negative. Note that, with all the extra positions, the total width becomes ten positions. Three of the ten positions are the decimal positions and the other seven are positions that are always needed for an E format. Thus, the total width of an E specification must be at least $d+7$; otherwise, asterisks will be printed. The specification above is E10.3.

If there are more decimal positions in the specification than are in the exponential form of the value, the extra decimal positions are filled on the right with zeros. If the total width of the E specification is more than 7 plus the decimal positions, the extra positions appear as blanks on the left side of the value.

EXAMPLE 2-10   ## TIME in an Exponential Form

Print the value of TIME in an exponential form with four decimal positions.

SOLUTION

Computer Memory

TIME   **− 0.00125**

FORTRAN Statements

```
      PRINT 10, TIME
   10 FORMAT (1X,'TIME = ',E11.4)
```

Buffer Contents

```
bTIMEb=b-0.1250E-02
```

Computer Output

```
          111111111
 123456789012345678
```

```
TIME = -0.1250E-02
```

. . . .

Try It   Try this self-test to check your memory of some key points from Section 2-4. If you have any problems with the exercises, you should reread this section. The solutions are given at the end of this module.

1. What is printed by the following statements? Show the exact location of any blanks.

```
      REAL X
      X = -27.632
      PRINT 5, X
    5 FORMAT (1X,'X = ',F7.1,' DEGREES')
```

2. What is printed by the following statements? Show the exact location of any blanks.

```
      REAL DIST, VEL
      DIST = 28732.5
      VEL = -2.6
      PRINT 5, DIST, VEL
    5 FORMAT (1X,'DISTANCE = ',E10.3,
   +          5X,'VELOCITY = ',F5.2)
```

2-5 COMPLETE PROGRAMS

The PROGRAM statement identifies the beginning of a program and assigns the program name. The general form of this statement is

> PROGRAM *program name*

Like a variable name, the program name can be one to six characters, must begin with a letter, and contains only letters and digits. Some example PROGRAM statements are

```
PROGRAM TEST
PROGRAM COMPUT
PROGRAM SORT2
```

You may not use a variable in your program that has the same name as the program name. The PROGRAM statement is not required, but we recommend using it to clearly identify the beginning of a program.

The STOP statement signals the computer to terminate execution of the program. It can appear anywhere in the program that makes sense, and it can appear as often as necessary. For example, certain data values may not be valid, and you may want to stop executing the program if they occur. The general form of the STOP statement is

> STOP

The STOP statement is optional in most programs, as explained in the next paragraph.

The END statement identifies the physical end of a FORTRAN program for the compiler. Since the compiler stops translating statements when it reaches the END statement, every FORTRAN program must terminate with the END statement, whose general form is

> END

If a STOP statement does not immediately precede the END statement, the compiler automatically adds one. Therefore, a program may use both a STOP and an END at the end of the program, although the STOP is not necessary. In the programs in this module, we will not use the STOP statement when it immediately precedes the END statement.

Specification statements must precede *executable statements* in a program. Therefore, FORTRAN programs should have the following general structure:

> PROGRAM statement
> Specification statements
> Executable statements
> END statement

If you look back at the program in Chapter 1, you will be able to identify the different groups of statements.

In addition to learning to write correct programs, it is important to learn to write programs with good programming style, that is, to write programs with a simple and consistent form. The following are some of the style and technique guidelines that we will follow in the example programs in this module:

1. We will mark the beginning and end of each program with a comment line containing a series of dashes. A PROGRAM statement will always be the first FORTRAN statement in a program.
2. A brief discussion of the purpose of the program will follow the PROGRAM statement. This information should be similar to the information in the first two steps of our problem-solving process — the problem statement and a description of the input and output.
3. Specification statements will be used to specify the type of all variables used in the program.
4. In longer programs, comment lines containing blanks will be used to separate portions of programs into sections that correspond to the decomposition of the problem solution.
5. Any statement number will be right-justified within the group of columns 1 through 5.

EXAMPLE 2-11   Convert Kilowatt-Hours to Joules

Write a program to convert an amount in kwh (kilowatt-hours) to joules. The amount in kwh is to be entered from the terminal. The output should be the number read in kwh and the converted value in joules. (Use the following conversion factor: joules = 3.6E+06 × kwh.)

SOLUTION

Step 1 is to state the problem clearly: Convert a value in kilowatt-hours to joules.

Step 2 is to describe the input and output:

Input — value in kilowatt-hours
Output — corresponding value in joules

Step 3 is to work a simple example by hand. Therefore, let the input value be 5.5 kilowatt-hours. Then the number of joules is 3.6E+06 × 5.5 = 20E+06.

Step 4 is to develop an algorithm. We start with the decomposition and then add more details to obtain the pseudocode. We can usually go directly from the pseudocode to the FORTRAN program.

Decomposition

Read amount in kilowatt-hours.
Convert amount to joules.
Print both amounts.

Pseudocode

Convert: Read kilowatt-hours
        joules ← 3.6E+06 × kilowatt-hours
        Print kilowatt-hours, joules

FORTRAN Program

```
*------------------------------------------------------------------*
      PROGRAM CONVRT
*
*  This program reads an input value in kilowatt-hours, and
*  then converts it to joules.  Both values are then printed.
*
      REAL KWH, JOULES
*
      PRINT*, 'ENTER ENERGY IN KILOWATT-HOURS'
      READ*, KWH
*
      JOULES = 3.6E+06*KWH
*
      PRINT 5, KWH, JOULES
    5 FORMAT (1X,F6.2,' KILOWATT-HOURS = ',E9.2,' JOULES')
*
      END
*------------------------------------------------------------------*
```

Step 5 is to test the program. The following computer output illustrates a sample run of the program using the same data that we used in the hand example.

```
ENTER ENERGY IN KILOWATT-HOURS
5.5
  5.50 KILOWATT-HOURS =  0.20E+08 JOULES
```

. . . .

In the program in Example 2–11, we printed a message to the user that specified the input that the program was expecting. After converting the value in kilowatt-hours to joules, we printed the value of kilowatt-hours along with the converted value in joules. This interaction between the program and the user resembles a conversation and is called *conversational computing.*

2-6 Application   **BACTERIA GROWTH**

Biology

A biology laboratory experiment involves the analysis of a strain of bacteria. Because the growth of bacteria in the colony can be modeled with an exponential equation, we are going to write a computer program to predict how many bacteria will be in the colony after a specified amount of time. Suppose that, for this type of bacteria, the equation to predict growth is

$$y_{new} = y_{old}e^{1.386t}$$

where y_{new} is the new number of bacteria in the colony, y_{old} is the initial number of bacteria in the colony, and t is the elapsed time in hours. Thus, when $t = 0$, we have

$$y_{new} = y_{old}e^{1.386 \cdot 0} = y_{old}$$

1. Problem Statement

Using the equation

$$y_{new} = y_{old}e^{1.386t}$$

predict the number of bacteria (y_{new}) in a bacteria colony given the initial number in the colony (y_{old}) and the time elapsed (t) in hours.

2. Input/Output Description

Input—the initial number of bacteria and the time elapsed
Output—the number of bacteria in the colony after the elapsed time

3. Hand Example

You will need your calculator for these calculations. For $t = 1$ hour and $y_{old} = 1$ bacterium, the new colony contains

$$y_{new} = 1 \cdot e^{1.386 \cdot 1} = 4.00$$

After 6 hours, the size of the colony is

$$y_{new} = 1 \cdot e^{1.386 \cdot 6} = 4088.77$$

If we start with 2 bacteria, after 6 hours the size of the colony is

$$y_{new} = 2 \cdot e^{1.386 \cdot 6} = 8177.54$$

4. Algorithm Development

Decomposition

Read y_{old}, t.
Compute y_{new}.
Print y_{old}, t, y_{new}.

Pseudocode

$$\text{Growth: Read } y_{\text{old}}, t$$
$$y_{\text{new}} \leftarrow y_{\text{old}} e^{1.386t}$$
$$\text{Print } y_{\text{old}}, t, y_{\text{new}}$$

FORTRAN Program

```
*----------------------------------------------------------------*
      PROGRAM GROWTH
*
*  This program reads the initial population and the time
*  elapsed and then computes and prints the predicted
*  population.
*
      REAL YOLD, YNEW, TIME
*
      PRINT*, 'ENTER INITIAL POPULATION'
      READ*, YOLD
      PRINT*, 'ENTER TIME ELAPSED IN HOURS'
      READ*, TIME
*
      YNEW = YOLD*EXP(1.386*TIME)
*
      PRINT 10, YOLD
   10 FORMAT (1X,'INITIAL POPULATION = ',F9.4)
      PRINT 20, TIME
   20 FORMAT (1X,'TIME ELAPSED (HOURS) = ',F9.4)
      PRINT 30, YNEW
   30 FORMAT (1X,'PREDICTED POPULATION = ',F9.4)
*
      END
*----------------------------------------------------------------*
```

This program and all other programs from the Application sections are available on a diskette that is available to your instructor. Check with your instructor to see if this information has been stored on your computer system so that you can access these programs.

5. Testing

The program output using data from one of the hand examples is

```
ENTER INITIAL POPULATION
1.0
ENTER TIME ELAPSED IN HOURS
6.0
INITIAL POPULATION =    1.0000
TIME ELAPSED (HOURS) =    6.0000
PREDICTED POPULATION = 4088.7722
```

Very large values of TIME will result in an overflow error. Negative values of TIME represent population decreases, but if the population falls below 1 the model is no longer applicable.

2-7 Application   STRIDE ESTIMATION

Mechanical Engineering

The physics that govern the motion of a simple pendulum enable us to estimate the time that it takes for a natural stride of a person given only the length of his or her leg. This is done by modeling the leg by a long, thin rod of uniform cross section that pivots about its upper end, as shown in the following diagram.

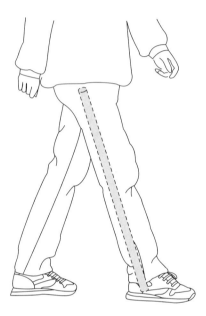

Using this model, a freely swinging rod supported at its upper end will take T seconds to swing from one end to another, and back to its original position, where[1]

$$T = 2 \cdot \pi \sqrt{\frac{2L}{3g}}$$

where L is the length of the rod (or leg) in feet and g is the acceleration due to gravity, which is approximately 32 ft/s². A stride will take $T/2$ seconds, since a stride is a step with only one leg.

It is interesting to observe that the length of a stride L_s is proportional to L, the length of the leg. Thus, we can express L_s as a product of a constant a and L:

$$L_s = aL$$

If we know the length of a stride and the time it takes to take a stride, we can compute the walking speed, which is the number of feet per second

---

[1] E. R. Jones and R. L. Childer, *Contemporary College Physics* (Reading, Mass.: Addison-Wesley, 1990), pp. 392–393.

traveled while walking:

$$\text{walking speed} = \frac{L_s}{\dfrac{T}{2}}$$

$$= \frac{aL}{\pi\sqrt{\dfrac{2L}{3g}}}$$

$$= \frac{a}{\pi\sqrt{\dfrac{2}{3g}}} \cdot \sqrt{L}$$

Therefore, the walking speed in feet per second is proportional to the square root of the length of the leg. Thus people with longer legs have a faster walking speed or go farther per second. This model does not work for running because the leg does not swing freely in a running gait.

Write a program that reads the length of a leg in feet, and estimates the time required for a stride.

1. Problem Statement

Compute the time required for a stride.

2. Input/Output Description

Input—the length of the leg
Output—the time required for a stride

3. Hand Example

Let the leg length be 3 feet. Then T_s, the time for a stride, is

$$T_s = \frac{T}{2} = \pi\sqrt{\frac{2L}{3g}} = \pi\sqrt{\frac{6}{96}} = 0.79 \text{ s}$$

Just for fun, measure your leg (from the hip joint to the heel) and use a stopwatch to time your natural walking stride. Compare that to the value computed using the previous equation. (It would also be interesting to measure the length of your stride and compute an estimate for the constant a. Compare your estimate for a with estimates from classmates with a wide range of heights. They should be similar.)

4. Algorithm Development

Decomposition

Read leg length.
Compute stride time.
Print stride time.

Pseudocode

Stride: Read L

$$T_s \leftarrow \pi \sqrt{\frac{2L}{3g}}$$

Print T_s

The FORTRAN program that follows defines the constants as separate variables so that the equation for T_s is as close as possible to the original equation. This makes the program easier to read and reduces the possibility of errors in converting the equation into FORTRAN.

FORTRAN Program

```
*-----------------------------------------------------------*
      PROGRAM STRIDE
*
*  This program reads the length of a leg, and then
*  computes and prints the time required for a stride.
*
      REAL LEG, PI, G, TIME
      PARAMETER (PI=3.141593, G=32.0)
*
      PRINT*, 'ENTER THE LEG LENGTH IN FEET'
      READ*, LEG
*
      TIME = PI*SQRT(2.0*LEG/(3.0*G))
*
      PRINT*
      PRINT 5, TIME
    5 FORMAT (1X,'THE STRIDE TIME IS ',F5.2,' SECONDS')
      PRINT 6, LEG
    6 FORMAT (1X,'FOR A LEG LENGTH OF ',F5.2,' FEET')
*
      END
*-----------------------------------------------------------*
```

5. Testing

If we test this program with the hand example data, the output is the following:

```
ENTER THE LEG LENGTH IN FEET
3

THE STRIDE TIME IS  0.79 SECONDS
FOR A LEG LENGTH OF  3.00 FEET
```

2-8 ADDITIONAL FORMATTING FEATURES

This section presents several useful features of FORMAT statements. Examples illustrate each feature.

Repetition

If we have two identical specifications in a row, we can use a constant in front of the specification (or sets of specifications) to indicate repetition. For instance, I2, I2, I2 can be replaced by 3I2. Often FORMAT statements can be made shorter with repetition constants. The following pairs of FORMAT statements illustrate the use of repetition constants:

```
10   FORMAT (3X,I2,3X,I2)
10   FORMAT (2(3X,I2))

20   FORMAT (1X,F4.1,F4.1,1X,I3,1X,I3,1X,I3)
20   FORMAT (1X,2F4.1,3(1X,I3))
```

Slash

A slash (/) in a FORMAT statement specifies that the current buffer should be printed and a new one started. The slash is especially useful in inserting blank lines in the output. However, do not assume that a slash will always cause single spacing; the spacing following the line printed by the slash depends on the carriage control character of the next line. The slash character may be enclosed in commas if desired.

The following statements print the heading TEST RESULTS followed by column headings TIME and HEIGHT:

```
PRINT 5
5 FORMAT (1X,'  TEST RESULTS'/1X,'TIME',5X,'HEIGHT')
```

The columns are separated by five spaces, and the heading is centered over the column headings, as shown below:

```
TEST RESULTS
TIME      HEIGHT
```

If we add another slash in the FORMAT statement, we have these statements:

```
PRINT 5
5 FORMAT (1X,'  TEST RESULTS'//1X,'TIME',5X,'HEIGHT')
```

The execution of these statements gives a blank line between the two headings:

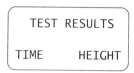

```
TEST  RESULTS

TIME     HEIGHT
```

Tab Specification

The tab specification Tn allows you to shift directly to a specified position, *n*, in the output line. The following pairs of FORMAT statements function exactly the same:

```
500   FORMAT (58X,'EXPERIMENT  NO.  1')
500   FORMAT (T59,'EXPERIMENT  NO.  1')

550   FORMAT (1X,'SALES',10X,'PROFIT',10X,'LOSS')
550   FORMAT (1X,'SALES',T17,'PROFIT',T33,'LOSS')

600   FORMAT (F6.1,15X,I7)
600   FORMAT (F6.1,T22,I7)
```

The TLn and TRn specifications tab left or right for n positions from the current position. The following formats are therefore equivalent:

```
85   FORMAT (1X,25X,'HEIGHT',5X,'WEIGHT')
85   FORMAT (T27,'HEIGHT',TR5,'WEIGHT')
```

The tab specifications are especially useful in aligning column headings and data.

Number of Specifications

Suppose there are more FORMAT specifications than variables on a PRINT list, as shown:

```
PRINT 1, SPEED, DIST
1 FORMAT (4F5.2)
```

In these cases the computer uses as much of the specification list as it needs and ignores the rest. In the example SPEED and DIST would be matched to the first two specifications; the last two specifications would be ignored.

Suppose there are fewer FORMAT specifications than variables on a PRINT list, as shown:

```
PRINT 20, TEMP, VOL
20 FORMAT (1X,F6.2)
```

In these cases we match variables and specifications until we reach the end of the FORMAT. Then, two events occur:

1. We print the current buffer and start a new one.
2. We back up in the FORMAT specification list until we reach a left parenthesis, and we again begin matching the remaining variables to the speci-

fications at that point. If a repetition constant is in front of this left parenthesis, it applies to the FORMAT specifications being reused.

In the previous statements TEMP would be matched to the F6.2 specification. Because there is no specification for VOL, we do the following:

1. Print the value of TEMP after single spacing.
2. Back up to the beginning of the FORMAT specification list (first left parenthesis) and match the F6.2 to the value of VOL. We then reach the end of the list and single space to print the value of VOL. TEMP and VOL are thus printed on separate lines.

This discussion allows you to understand what happens if the number of specifications is not the same as the number of variables. However, we recommend that you reference the same FORMAT with more than one PRINT statement only when the FORMAT specifications are exactly the same. Write a new FORMAT statement if the number of variables is different; this keeps the program simpler to understand.

Try It

Try this self-test to check your memory of some key points from Section 2–8. If you have any problems with the exercises, you should reread this section. The solutions are included at the end of this module.

In problems 1–3, show the output from the following PRINT statements. Be sure to indicate the vertical as well as the horizontal spacing. Use the following variables and corresponding values:

$$TIME = 3.5, \quad RESP1 = 178.8, \quad RESP2 = 0.00204$$

```
1.   PRINT 5, TIME, RESP1, RESP2
     5 FORMAT (1X,F6.2,T20,2F7.4)
2.   PRINT 4, TIME, RESP1, TIME, RESP2
     4 FORMAT (1X,'TIME = ',F5.2,TR5,'RESPONSE 1 = ',E9.2/
    +          1X,'TIME = ',F5.2,TR5,'RESPONSE 2 = ',E9.2)
3.   PRINT 1, TIME, RESP1, RESP2
     1 FORMAT (1X,'EXPERIMENT RESULTS'//
    +          1X,'TIME',2X,'RESPONSE 1',2X,'RESPONSE 2'/
    +          1X,F4.2,2F12.3)
```

In problems 4–7, tell how many data lines are printed by the following PRINT statements. Indicate which variables are on each line and which columns are used.

```
4.    PRINT 4, TIME, DIST, VEL, ACCEL
      4 FORMAT (4(1X,F6.3))
5.    PRINT 14, TIME, DIST, VEL, ACCEL
     14 FORMAT (1X,F6.2)
6.    PRINT 2, TIME, DIST, VEL, ACCEL
      2 FORMAT (1X,F4.2/1X,F4.1)
7.    PRINT 3, TIME, DIST
      PRINT 3, VEL, ACCEL
      3 FORMAT (1X,4F6.3)
```

RY

This chapter discussed how to define variables and constants in FORTRAN. It discussed the arithmetic operations and intrinsic functions that allow us to compute new values using these variables and constants. Some of the considerations that are unique to computer computations were discussed with specific examples: magnitude limitations, truncation, mixed-mode operations, underflow, and overflow. Statements for reading information from the terminal and for printing answers were covered. Several complete example programs were developed.

Key Words

argument	intrinsic function
arithmetic expression	left-justified
assignment statement	list-directed I/O
buffer	literal
carriage control character	mixed-mode operation
comment line	nonexecutable statement
constant	overflow
continuation line	real value
conversational computing	right-justified
executable statement	rounding
explicit typing	scientific notation
exponential notation	specification statement
formatted I/O	statement numbers
generic function	truncation
implicit typing	type statement
initialize	underflow
integer value	variable
intermediate result	

Problems

This problem set begins with modifications to programs given earlier in this chapter. Give the decomposition, pseudocode, and FORTRAN program for each problem.

Problem 1 modifies the energy conversion program CONVRT given in Section 2-5.

1. Modify the energy conversion program so that it converts kilowatt-hours to calories. (1 calorie equals 4.19 joules)

Problem 2 modifies the bacteria growth program GROWTH given in Section 2-6.

2. Modify the bacteria growth program so that the program reads two time values from the terminal, with no restrictions on which time is larger. Compute and print the amount of growth between the two times, assuming an initial population value of 1. (*Hint:* Review the absolute value function.)

Problem 3 modifies the stride estimation program STRIDE given in Section 2-7.

3. Experimentally determine a value for the constant a. Then modify the stride estimation program so that it accepts the leg length and computes the time required for a stride, the length of a stride, and the walking speed.

Develop programs for problems 4-9. Use the five-phase design process.

4. Hot-air balloons act on the principle that heated air is less dense than cool air. The mass of air inside the balloon, for a fixed volume, is less than the mass of an equivalent volume of cooler air outside. This occurs because as the air inside is heated, it expands and some escapes in order to maintain constant atmospheric pressure. Thus, the air inside is less dense than the air outside and the balloon floats. Write a program to determine the mass of air that remains (m_2) after reading the original mass of air (m_1), the original volume (v_1), and the volume after heating (v_2). Use the following equation for computing m_2:

$$m_2 = \frac{v_1}{v_2} m_1$$

5. When a train travels over a straight section of track, it exerts a downward force on the rails; but when it rounds a level curve, it also exerts a horizontal force outward on the rails. Both of these forces must be considered when designing the track. The downward force is equivalent to the weight of the train. The horizontal force, called centrifugal force, is a function of the weight of the train, its speed as it rounds the curve, and the radius of the curve. The equation to compute the horizontal force in pounds is

$$\text{force} = \frac{\text{weight} \cdot 2000}{32} \cdot \frac{(\text{mph} \cdot 1.4667)^2}{\text{radius}}$$

where weight is the weight of the train in tons, mph is the speed of the train in miles per hour, and radius is the radius of the curve in feet. Write a program to read values for weight, mph, and radius. Compute and print the corresponding horizontal force generated.

6. Modify the program in Problem 5 so that the speed is entered in kilometers per hour instead of miles per hour. (Recall that 1 mile is equal to 1.609 kilometers.)

7. A research scientist performed nutrition tests using three animals. Data on each animal include an identification number, the weight of the animal at the beginning of the experiment, and the weight of the animal at the end of the experiment. Write a program to read this data and print a report. The report is to include the original information plus the percentage increase in weight for each animal.

8. Write a program to read the following information from the terminal:

> Year
> Number of people in the civilian labor force
> Number of people in the military labor force

Compute the percentage of the labor force that is civilian and the percentage that is military. Print the following information:

```
LABOR FORCE  - YEAR XXXX
NUMBER OF WORKERS (THOUSANDS) AND PERCENTAGE OF WORKERS
CIVILIAN        XXX.XXX      XXX.XXX
MILITARY        XXX.XXX      XXX.XXX
TOTAL           XXX.XXX      XXX.XXX
```

9. The approximate time for electrons to travel from the cathode to the anode of a rectifier tube is given by

$$\text{TIME} = \sqrt{\frac{2M}{Q \cdot V}} \cdot R1 \cdot Z \cdot \left(1 + \frac{Z}{3} + \frac{Z^2}{10} + \frac{Z^3}{42} + \frac{Z^4}{216} \right)$$

where Q = charge of the electron ($1.60206E-19$ coulombs)
   M = mass of the electron ($9.1083E-31$ kilograms)
   V = accelerating voltage in volts
   $R1$ = radius of the inner tube (cathode)
   $R2$ = radius of the outer tube (anode)
   Z = natural logarithm of $R2/R1$

Define the values of Q and M in the constant section of your program. Read values for V, $R1$, and $R2$, then calculate Z and TIME. Print the values of V, $R1$, $R2$, and TIME.

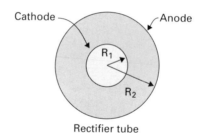

Rectifier tube

3 Control Structures

An optical fiber (a transparent glass thread) is thinner than a human hair, but it can carry more information than either radio waves or electrical waves in copper telephone wires. In addition, fiber optic communication signals do not produce electromagnetic waves that cause "cross-talk" noise on communication lines. The first transoceanic fiber optic cable was laid in 1988 across the Atlantic. It contains four fibers that can handle up to 40,000 calls at one time. Other applications for optical fibers include motion sensing in gyroscopes, linking industrial lasers to machining tools, and threading light into the human body for examinations and laser surgery.

INTRODUCTION

In Chapter 2 we wrote complete FORTRAN programs, but the steps were all executed sequentially. The programs were composed of reading data, computing new data, and printing the new data. This chapter introduces FORTRAN statements that allow us to control the sequence of the steps being executed. This control is achieved through statements that allow us to select different paths through our programs and statements that allow us to repeat certain parts of our programs. These new statements are used to implement *control structures*.

3-1 ALGORITHM STRUCTURE

Chapter 1 presented the following five-step process for developing problem solutions:

1. State the problem clearly.
2. Describe the input and the output.
3. Work the problem by hand (or with a calculator) for a simple set of data.
4. Develop an algorithm that is general in nature.
5. Test the algorithm with a variety of data sets.

To describe algorithms consistently, we use a set of standard forms, or structures. When an algorithm is described in these standard structures, it is a *structured algorithm*. When the algorithm is converted into computer instructions, the corresponding program is a *structured program*. This chapter begins with a discussion of pseudocode and flowcharts, which describe the steps in an algorithm; it then discusses organizing the steps in standard structures.

Pseudocode and Flowcharts

Each structure for building algorithms can be described in an English-like notation called *pseudocode*. Because pseudocode is not really computer code, it is language independent; that is, pseudocode depends only on the steps needed to solve a problem, not on the computer language that will be used for writing the solution. *Flowcharts* also describe the steps in algorithms, but they are a graphical description, as opposed to a set of English-like steps. Neither pseudocode nor flowcharts are intended to be a formal way of describing the algorithm; they are informal ways of easily describing the steps in the algorithm without worrying about the syntax of a specific computer language. This chapter includes a number of examples of both pseudocode and flowcharts so that you can compare the techniques and choose the one that works best for you.

The basic operations that we perform in algorithms are computations, input, output, and comparisons. Table 3-1 compares examples of pseudocode notation and flowchart symbols for these basic operations, plus the notation and symbols for identifying the beginning and end of an algorithm. Do not worry about whether the names you have chosen are valid FORTRAN names; both pseudocode and flowcharts are describing the operations, and they are independent of the language that will be used to implement the computer solution.

Table 3–1 Pseudocode Notation and Flowchart Symbols

Basic Operation	Pseudocode Notation	Flowchart Symbol
Computations	average $\leftarrow \dfrac{sum}{count}$	$average = \dfrac{sum}{count}$
Input	Read A, B	Read A, B
Output	Print A, B	Print A, B
Comparisons	If A > 0.0 then . . .	Is A > 0.0? yes / no
Beginning of algorithm	Report:	Start Report
End of algorithm		Stop Report

In pseudocode a computation is indicated with an arrow, as shown in the following statement:

$$average \leftarrow \frac{sum}{count}$$

We read this pseudocode statement as "the average is replaced by the sum divided by the count" or "the sum is divided by the count, giving the average."

We use comparison to ask a question within an algorithm. Then, depending on the answer to the question, we can perform one set of statements as opposed to another set of statements. The questions that we ask must be answered with yes or no (or true or false). For example, we can determine if a data value A is greater than zero by using the expressions "If $A > 0.0$, then . . ." or "Is $A > 0.0$?"

An algorithm defined in pseudocode begins with the name of the algorithm, followed by the first step in the algorithm. Thus, if the first step in an algorithm named Report is to set a sum to zero, the first pseudocode statement is

$$Report: sum \leftarrow 0$$

Pseudocode does not need a special statement to signify the end of the algorithm; we assume that the end occurs after the last step described in pseudocode.

General Structures

The steps in an algorithm can be divided into three general structures — sequences, selections, and repetition. A sequence is a set of steps that are performed sequentially, or one after another. A selection structure allows us to compare two values and then specify a different set of steps to execute, depending on the result of the comparison. We use the repetition structure when we want to repeat a set of steps. We can repeat the steps a specified number of times, or we can repeat the steps as long as a specified condition is true. Now let us look at each of these structures separately.

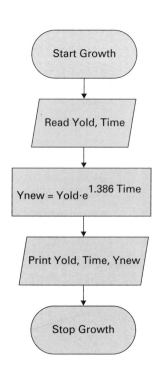

Sequence A *sequence* is a set of steps in an algorithm that are performed sequentially. For example, all of the programs in Chapter 2 were sequential algorithms that contained an input step, a computation step, and an output step. We included sample pseudocode in the development of the programs in Chapter 2; shown opposite is the flowchart for the program GROWTH from Section 2-6.

Selection In the *selection* structure, a comparison is performed to determine which steps are to be performed next. The selection structure is commonly described in terms of an If structure that can have several forms.

All forms of the structure include a comparison that tests a *condition* that can be evaluated to be true or false. If the condition is true, then a step or a group of steps is performed. For example, suppose the algorithm that we are developing must count the number of students who are on the honor roll and print their student ID numbers. One step in the algorithm might be "if the grade point average (GPA) is greater than or equal to 3.0, then print the ID and add 1 to the number of students on the honor roll." This step is shown in both pseudocode and a flowchart; note the use of indenting in the pseudocode.

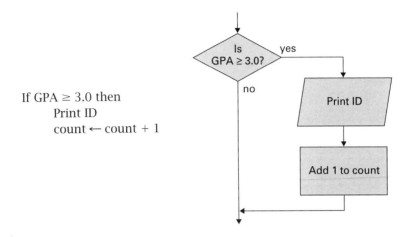

If GPA ≥ 3.0 then
    Print ID
    count ← count + 1

Another form of the If structure contains an additional clause called an Else clause that specifies an alternate set of steps to perform if the condition is false. For example, suppose we want to modify the example so that we print the ID of each student. Then, if the student is on the honor roll, we also print the GPA on the same line beside the student's ID. These steps are described with the following pseudocode and flowchart:

If GPA ≥ 3.0 then
    Print ID, GPA
    count ← count + 1
Else
    Print ID

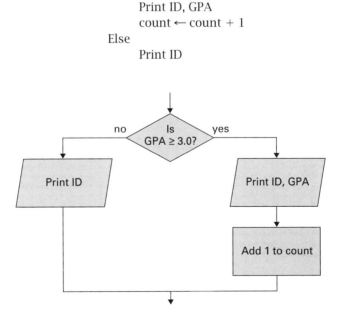

The last form of the If structure contains Else If clauses that allow us to test for multiple conditions. A single Else If clause can be illustrated by extending the honor roll example. Assume that the president's honor roll requires a GPA greater than 3.5 and that the dean's honor roll requires a GPA greater than or equal to 3.0. We want to count the number of students on each honor roll. We also want to print each student's ID and GPA if the GPA is above 3.0; in addition, we want to include an asterisk beside the GPA if it is above 3.5. The following pseudocode and flowchart describe this set of steps:

> If GPA > 3.5 then
> > Print ID, GPA, '*'
> > president's count ← president's count + 1
> Else If GPA ≥ 3.0 then
> > Print ID, GPA
> > dean's count ← dean's count + 1
> Else
> > Print ID

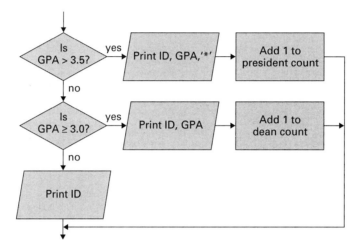

Repetition A *repetition* structure allows us to use *loops*, which are sets of steps in an algorithm that are repeated. One type of loop, called a While loop, repeats the steps as long as a certain condition is true. Another type of loop, called a *counting loop*, repeats the steps a specified number of times. The steps within a loop can contain sequential steps, selection steps, or other repetition steps.

Suppose that we wish to continue reading data values and adding them to a sum as long as that sum is less than 1000. These steps are described in the following pseudocode and flowchart:

While sum < 1000 do
> Read data value
> sum ← sum + data value

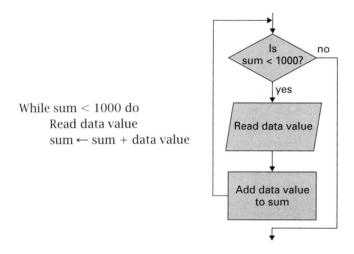

Again, note that the indenting in the pseudocode specifies the steps that are included in the While loop. If we wish to print the value of the sum after exiting the While loop, we can use the following pseudocode and flowchart:

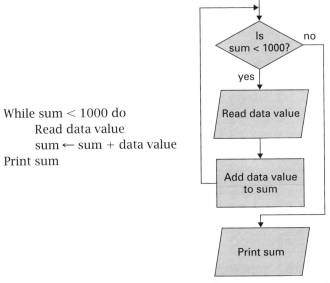

While sum < 1000 do
      Read data value
      sum ← sum + data value
Print sum

Counting loops repeat a set of steps a specified number of times. For example, suppose we are going to read ten values from the terminal and perform some calculations with each value. We can describe the steps in a counting loop that is repeated ten times. However, a counting loop can also be considered a special form of a While loop in which a counter has been introduced. The counter represents the number of times that the loop has been executed. The counter is usually initialized to zero before the loop is executed. Inside the loop, the counter is incremented by one. The loop is then executed "while the value of the counter is less then ten." The pseudocode and flowchart for this counting loop example are the following:

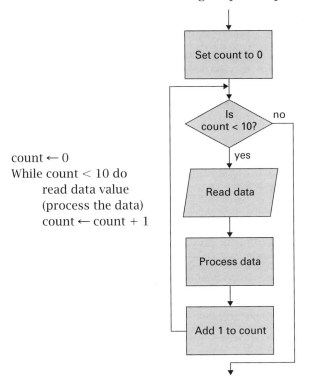

count ← 0
While count < 10 do
      read data value
      (process the data)
      count ← count + 1

We can describe algorithms for almost any problem using these three structures — sequential steps, selection steps, and repetitive steps.

3-2 IF STRUCTURES

Section 3-1 presented the pseudocode and flowcharts for the three forms of the selection structure. This section presents the corresponding FORTRAN statements. All three forms used an *If structure,* which contains a logical expression that is evaluated to determine which path to take in the structure. Therefore, before discussing the FORTRAN statements, this section covers logical expressions that are used as the logical expressions to be tested.

Logical Expressions

A *logical expression* is analogous to an arithmetic expression but is always evaluated to either true or false, instead of a number. Logical expressions can be formed using the following *relational operators:*

Relational Operator	Interpretation
.EQ.	equal to
.NE.	not equal to
.LT.	less than
.LE.	less than or equal to
.GT.	greater than
.GE.	greater than or equal to

Numeric variables can be used on both sides of the relational operators to yield a logical expression whose value is either true or false. For example, consider the logical expression A.EQ.B, where A and B are real values. If the value of A is equal to the value of B, then the logical expression A.EQ.B is true. Otherwise, the expression is false. Similarly, if the value of X is 4.5, then the expression X.GT.3.0 is true.

We can also combine two logical expressions into a *compound logical expression* with the *logical operators* .OR. and .AND. When two logical expressions are joined by .OR., the entire expression is true if either or both expressions are true; it is false only when both expressions are false. When two logical expressions are joined by .AND., the entire expression is true only if both expressions are true. These logical operators are used only between complete logical expressions. For example, A.LT.B.OR.A.LT.C is a valid compound logical expression because .OR. joins A.LT.B and A.LT.C. However, A.LT.B.OR.C is an invalid compound expression because C is a numeric variable, not a complete logical expression.

Logical expressions can also be preceded by the logical operator NOT. This operator changes the value of the expression to the opposite value; hence, if A.GT.B is true, then .NOT.A.GT.B is false.

A logical expression may contain several logical operators, as in

```
.NOT.(A.LT.15.4).OR.KT.EQ.ISUM
```

The *hierarchy,* from highest to lowest, is .NOT., .AND., and .OR. In the pre-

Table 3-2   Logical Operators

A	B	.NOT.A	A.AND.B	A.OR.B	A.EQV.B	A.NEQV.B
False	False	True	False	False	True	False
False	True	True	False	True	False	True
True	False	False	False	True	False	True
True	True	False	True	True	True	False

ceding statement the logical expression A.LT.15.4 would be evaluated, and its value, true or false, would then be reversed. This resultant value would be used, along with the value of KT.EQ.ISUM, with the logical operator .OR. For example, if A is 5.0, KT is 5, and ISUM is 5, then the left-hand expression is false and the right-hand expression is true; but these expressions are connected by .OR., thus the entire expression is true.

Another type of variable, a *logical variable,* is also useful in writing the logical expressions that allow us to choose different paths and to repeat parts of our programs. A logical variable can have one of two values: true or false. A logical variable must be defined with a specification statement whose form is

LOGICAL *variable list*

Logical constants are .TRUE. and .FALSE. Therefore, the statements necessary to define a logical variable and give it a value of false are

```
LOGICAL DONE
DONE = .FALSE.
```

Logical variables are generally used to make programs more readable; thus, they are not usually part of the input or output. Several examples in this chapter will compare solutions using logical variables to ones without logical variables.

It is invalid to compare two logical variables with the relation .EQ. or .NE. Instead, two new relations, .EQV. and .NEQV., are used to represent equivalent and not equivalent. To compare .NOT.DONE to the value .TRUE., we could use this statement:

```
IF (.NOT.DONE.EQV..TRUE.) THEN
```

Table 3-2 summarizes the evaluation of the logical operators for all possible cases.

Whenever arithmetic, relational, and logical operators are in the same expression, the arithmetic operations are performed first; the relational operators are then applied to yield true or false values; and these values are evaluated with the logical operators, whose precedence is .NOT., .AND., and .OR. The relations .EQV. and .NEQV. are evaluated last.

As an example, assume that ERROR is a logical variable and that A, B, and C are real variables. Consider the following expression:

```
A.GT.B+C.OR.ERROR
```

Table 3-3   Relational and Arithmetic Operator Precedence

Priority	Operation	Order
1.	Parentheses	Innermost first
2.	Exponentiation	Right to left
3.	Multiplication and division	Left to right
4.	Addition and subtraction	Left to right (unary first, as in $-A$, then binary, as in $A - B$)
5.	Relational operators	Left to right
6.	.NOT.	Left to right
7.	.AND.	Left to right
8.	.OR.	Left to right
9.	.EQV., .NEQV.	Left to right

This expression will be true if either $A > B + C$ or ERROR is .TRUE.; the expression will also be true if both $A > B + C$ and ERROR is .TRUE. Table 3-3 lists the precedence of operators in a logical expression.

Logical IF Statement

FORTRAN implements a logical IF statement and a block IF statement. The logical IF statement is used if a single statement is to be performed if a logical expression is true. We use the block IF statement if we want to perform several statements if a logical expression is true, or if we want to use an ELSE statement, or if we want to use an ELSE IF statement.

The general form of the logical IF statement is

> IF *(logical expression) executable statement*

Execution of this statement consists of the following steps:

1. If the logical expression is true, we execute the statement that is on the same line as the logical expression and then go to the next statement in the program.
2. If the logical expression is false, we jump immediately to the next statement in the program.

A typical IF statement is the following:

```
IF (A.LT.B) SUM = SUM + A
```

If the value of A is less than the value of B, then the value of A is added to SUM. If the value of A is greater than or equal to B, then control passes to whatever statement follows the IF statement in the program. Other examples of IF statements are

```
IF (TIME.GT.1.5) READ*, DISTNC
IF (DEN.LE.0.0) PRINT*, DEN
IF (-4.NE.NUM) NUM = NUM + 1
```

The executable statement that follows the logical expression is typically a computation or an input/output statement—it cannot be another IF statement.

Block IF Statement

In many instances we would like to perform more than one statement if a logical expression is true. The form of the IF statement that allows us to perform any number of statements if a logical expression is true uses the words THEN and END IF to identify these steps. The general form is

```
IF (logical expression) THEN
        statement 1
            .
            .
            .
        statement n
END IF
```

Execution of this block of statements consists of the following steps:

1. If the logical expression is true, we execute statements 1 through *n* and then go to the statement following END IF.
2. If the logical expression is false, we jump immediately to the statement following END IF.

Although not required, indenting the statements to be performed when the logical expression is true will indicate that they are a group of statements within the IF statement.

EXAMPLE 3-1

Zero Divide

Assume that you have calculated the numerator NUM (explicitly typed REAL) and the denominator DEN of a fraction. Before dividing the two values, you want to see if DEN is zero. If DEN is zero, you want to print an error message and stop the program; if DEN is not zero, you want to compute the result and print it. Write the statements to perform these steps.

SOLUTION

```
IF (DEN.EQ.0.0) THEN
    PRINT*, 'DENOMINATOR IS ZERO'
    STOP
END IF
FRACTN = NUM/DEN
PRINT*, 'FRACTION = ', FRACTN
```

. . . .

ELSE Statement The ELSE statement allows us to execute one set of statements if a logical expression is true and a different set if the logical expression is false. The general form of an IF statement combined with an ELSE statement (an IF-ELSE statement) is shown in the following box. If the logical expression is true, then statements 1 through *n* are executed. If the logical expression is false, then statements *n* + 1 through *m* are executed. Any statement can also be another IF or IF-ELSE statement to provide a nested structure.

```
IF (logical expression) THEN
       statement 1
           .
           .
           .
       statement n
ELSE
       statement n + 1
           .
           .
           .
       statement m
END IF
```

Now consider this set of statements that uses a logical variable VALID with the IF structure:

```
READ*, VOLTS
IF (VOLTS.GE.-5.0.AND.VOLTS.LE.5.0) THEN
   VALID = .TRUE.
ELSE
   VALID = .FALSE.
END IF
IF (.NOT.VALID) PRINT*, 'ERROR IN DATA'
```

In these statements, the variable VALID is true if the value in VOLTS is between −5.0 and 5.0; otherwise, VALID is false. After determining the value of VALID, it can be used in other places in the program, such as in an IF statement to print an error message if the value is not valid. The advantage of using the logical variable is that we can reference it each time we need to know if the value is valid, instead of repeating the test to see if the value in VOLTS is between −5.0 and 5.0.

ELSE IF Statement When we nest several levels of IF-ELSE statements, it may be difficult to determine which logical expressions must be true (or false) to execute a set of statements. In these cases we can use the ELSE IF statement to clarify the program logic. The general form of this statement is

```
IF (logical expression 1) THEN
    statement 1
        .
        .
        .
    statement m
ELSE IF (logical expression 2) THEN
    statement m + 1
        .
        .
        .
    statement n
ELSE IF (logical expression 3) THEN
    statement n + 1
        .
        .
        .
    statement p
ELSE
    statement p + 1
        .
        .
        .
    statement q
END IF
```

This general form has two ELSE IF statements; there may be more or less in an actual construction. If logical expression 1 is true, then only statements 1 through m are executed. If logical expression 1 is false and logical expression 2 is true, then only statements $m + 1$ through n are executed. If logical expressions 1 and 2 are false and logical expression 3 is true, then only statements $n + 1$ through p are executed. If more than one logical expression is true, the first true logical expression encountered is the only one executed.

If none of the logical expressions are true, then statements $p + 1$ through q are executed. If there is not a final ELSE statement and none of the logical expressions are true, then the entire construction is skipped. The IF-ELSE IF form is also called a CASE structure because a number of cases are tested. Each case is defined by its corresponding logical expression.

EXAMPLE 3-2

Weight Category

An analysis of a group of weight measurements involves converting a weight value into an integer category number that is determined as follows:

Category	Weight (pounds)
1	weight $\leq$ 50.0
2	50.0 < weight $\leq$ 125.0
3	125.0 < weight $\leq$ 200.0
4	200.0 < weight

Write FORTRAN statements that put the correct value (1, 2, 3, or 4) into CATEGR based on the value of WEIGHT. Assume that CATEGR has been explicitly typed as an integer variable.

SOLUTION 1

This solution uses nested IF-ELSE statements:

```
IF (WEIGHT.LE.50.0)THEN
    CATEGR = 1
ELSE
    IF (WEIGHT.LE.125.0)THEN
        CATEGR = 2
    ELSE
        IF (WEIGHT.LE.200.0)THEN
            CATEGR = 3
        ELSE
            CATEGR = 4
        END IF
    END IF
END IF
```

SOLUTION 2

This solution uses the IF-ELSE IF statement:

```
IF (WEIGHT.LE.50.0) THEN
    CATEGR = 1
ELSE IF (WEIGHT.LE.125.0) THEN
    CATEGR = 2
ELSE IF (WEIGHT.LE.200.0) THEN
    CATEGR = 3
ELSE
    CATEGR = 4
END IF
```

As you can see, Solution 2 is more compact than Solution 1: It combines the ELSE and IF statements into the single statement ELSE IF and eliminates two of the END IF statements.

The order of the logical expressions is important in Solution 2, because the evaluation will stop as soon as a true logical expression has been encountered. Changing the order of the logical expressions and the category assignments can cause the CATEGR value to be set incorrectly.

. . . .

Try It Try this self-test to check your memory of some key points from Section 3-2. If you have any problems with the exercises, you should reread this section. The solutions are given at the end of this module.

For problems 1–8, use the values given to determine whether the following logical expressions are true or false.

A = 2.2   B = −1.2   I = 5   DONE = .TRUE.

1. A.LT.B

2. A - B.GE.6.5

3. I.NE.5

4. A + B.GE.B

5. .NOT.(A.EQ.2*B)

6. I.LE.I - 5

7. (A.LT.10.0).AND.(B.GT.5.0)

8. (ABS(I).GT.2).OR.DONE

For problems 9–16, give FORTRAN statements that perform the steps indicated.

9. If TIME is greater than 5.0, increment TIME by 0.5.

10. When the square root of POLY is greater than or equal to 8.0, print the value of POLY.

11. If the difference between VOLT1 and VOLT2 is smaller than 6.0, print the values of VOLT1 and VOLT2.

12. If the absolute value of DEN is less than 0.005, print the message 'DENOMINATOR IS TOO SMALL.'

13. If the natural logarithm of X^2 is greater than or equal to 3, set TIME equal to zero and add X to SUM.

14. If DIST is less than 50.0 or TIME is greater than 10.0, increment TIME by 1.0. Otherwise, increment TIME by 0.5.

15. If DIST is greater than or equal to 50.0, increment TIME by 2.0 and print a message 'DISTANCE > 50.0.'

16. If DIST is greater than 100.0, increment TIME by 2.0. If DIST is between 50.0 and 100.0 (including 100.0), increment TIME by 1.0. Otherwise, increment TIME by 0.5.

3-3 Application   **LIGHT PIPES**

Optical Engineering

If we direct light into one end of a long rod of glass or plastic, the light is totally reflected by the walls, bouncing back and forth until it emerges at the far end of the rod. Light pipes use this interesting optical phenomenon to transmit light, and even images, from one place to another. If we bend a light pipe, the light will follow the shape of the pipe and emerge only at the end, as shown in the diagram below:

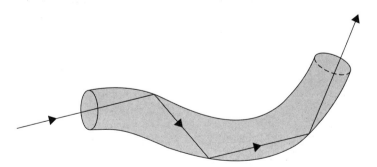

Light pipes made of very thin optical fibers can be grouped into bundles. If the fiber ends are polished and the spatial arrangement is the same at both ends (a coherent bundle), the fiber bundle can be used to transmit an image, and the bundle is called an image conduit. If the fibers do not have the same arrangement at both ends (an incoherent bundle), light is transmitted instead of an image, and the bundle is called a light guide. Because the optical fibers are so flexible, light guides and image conduits are used in instruments designed to permit visual observation of objects or areas that would otherwise be inaccessible. For example, an endoscope is an instrument used by physicians to examine the interior of a patient's body with only a very small incision.

This phenomenon of total internal reflection can be predicted using Snell's law and the indices of refraction of the materials being considered for the light pipe. A light pipe is actually composed of two materials: the material forming the rod or pipe itself and the material that surrounds the pipe. Normally, the material forming the rod is denser than the surrounding medium. When light passes from one material into another material with a different density, the light is bent, or refracted, at the interface of the two materials. The amount of refraction depends on the indices of refraction of the materials and the angle of incidence of the light. If the light striking the interface comes from within the denser material, it may reflect off the interface rather than pass through it. The angle of incidence where the light will be reflected from the surface, rather than cross it, is called the critical angle θ_c. Since the critical angle depends on the indices of refraction of the two materials, we can compute this angle and determine whether light entering the pipe at a particular angle will stay within the pipe. Assume n_2 is the index of refraction of the surrounding medium and n_1 is the index of refraction of the pipe itself. If n_2 is greater than n_1, the pipe will not transmit light; otherwise, the critical angle can be determined from the following equation:

$$\sin \theta_c = \frac{n_2}{n_1}$$

Write a program that reads the indices of refraction for two materials that form a pipe and the angle at which light enters the pipe. Print a message indicating if the pipe will transmit light or not.

1. Problem Statement

Determine whether a light pipe generated from two materials will transmit light that enters it at a given angle.

2. Input/Output Description

Input—indices of refraction of the two materials and the angle at which light enters the pipe

Output—message indicating whether or not light is transmitted

3. Hand Example

The index of refraction of air is 1.0003 and the index of glass is 1.5. If we form a light pipe of glass surrounded by air, the critical angle θ_c can be computed as follows:

$$\theta_c = \sin^{-1}\left(\frac{n_2}{n_1}\right) = \sin^{-1}\left(\frac{1.0003}{1.5}\right)$$

$$= \sin^{-1}(0.66687) = 41.82°$$

This light pipe will transmit light for all angles of incidence greater than 41.82°.

4. Algorithm Development

Decomposition

Read n_1, n_2, and incidence angle.
Determine appropriate message.

We will need to make sure that n_2 is not greater than n_1 before we compute the critical angle, since the inverse sine function will give an error message if its argument is greater than 1.0.

Pseudocode

Pipe: Read n_1, n_2, and incidence angle
      If $n_2 > n_1$ then
            Print 'Light is not transmitted'
      Else

            critical angle $\leftarrow \sin^{-1}\left(\dfrac{n_2}{n_1}\right)$

            If incidence angle > critical angle then
                 Print 'Light is transmitted'
            Else
                 Print 'Light is not transmitted'

As we convert the pseudocode into FORTRAN, we need to be sure to remind the user to use the proper units for the angle measurement. We can write the program to accept either radians or degrees, but the equation for computing the critical angle will depend on the units of the input angle.

FORTRAN Program

```
*---------------------------------------------------------------*
      PROGRAM PIPE
*
*  This program reads the indices of refraction for two
*  materials forming a light pipe. It also reads the angle of
*  incidence for light striking the pipe and determines if
*  it is transmitted.
```

```
*
      REAL N1, N2, ANGLE, CRTCL
*
      PRINT*, 'ENTER INDEX OF REFRACTION FOR ROD'
      READ*, N1
      PRINT*, 'ENTER INDEX OF REFRACTION FOR SURROUNDING MEDIUM'
      READ*, N2
      PRINT*, 'ENTER ANGLE OF TRANSMISSION OF LIGHT IN DEGREES'
      READ*, ANGLE
*
      IF (N2.GT.N1) THEN
         PRINT*, 'LIGHT IS NOT TRANSMITTED'
      ELSE
         CRTCL = ASIN(N2/N1)*(180.0/3.141593)
         IF (ANGLE.GT.CRTCL) THEN
            PRINT*, 'LIGHT IS TRANSMITTED'
         ELSE
            PRINT*, 'LIGHT IS NOT TRANSMITTED'
         END IF
      END IF
*
      END
*-------------------------------------------------------------*
```

5. Testing

To test this program with a variety of possible pipes, we can use the following list of indices of refraction for some common materials.[1]

Material	Index of Refraction
Gases (at atmospheric pressure and 0°C)	
Hydrogen	1.0001
Air	1.0003
Carbon dioxide	1.0005
Liquids (at 20°C)	
Water	1.333
Ethyl alcohol	1.362
Glycerine	1.473
Solids (at room temperature)	
Ice	1.31
Polystyrene	1.59
Crown glass	1.50–1.62
Flint glass	1.57–1.75
Diamond	2.417
Acrylic (polymethylmethacrylate)	1.49

[1] E. R. Jones and R. L. Childers, *Contemporary College Physics* (Reading, MA: Addison-Wesley, 1990).

If we use this program to determine whether light with an incidence angle of 45° is transmitted through an acrylic pipe in water, we get the following output:

```
ENTER INDEX OF REFRACTION FOR ROD
1.49
ENTER INDEX OF REFRACTION FOR SURROUNDING MEDIUM
1.333
ENTER ANGLE OF TRANSMISSION OF LIGHT IN DEGREES
45.0
LIGHT IS NOT TRANSMITTED
```

3-4   WHILE LOOP STRUCTURE

The *While loop* is an important structure for repeating a set of statements as long as a certain condition is true. In pseudocode, the While loop structure is

> While *condition* do
>   *statement 1*
>   .
>   .
>   .
>   *statement m*
>   *statement p*

While the condition is true, statements 1 through *m* are executed. After the group of statements is executed, the condition is retested. If the condition is still true, the group of statements is reexecuted. When the condition is false, execution continues with the statement following the While loop (statement *p* in our example). The variables modified in the group of statements in the While loop must involve the variables tested in the While loop's condition, or the value of the condition will never change.

Standard FORTRAN 77 does not include a WHILE statement, although many compilers have implemented their own WHILE statement. We will implement the While loop with the IF statement as shown below, so that our programs will be standard and will execute on other FORTRAN 77 compilers without any conversion.

```
n IF (logical expression) THEN
      statement 1
          .
          .
          .
        statement m
      GO TO n
   END IF
```

In this implementation we used an unconditional transfer statement whose general form is

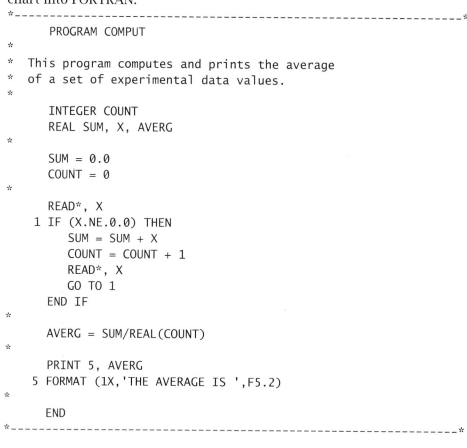

where *n* is the statement number or label of an executable statement in the program. The execution of the GO TO statement causes the flow of program control to transfer, or *branch*, to statement *n*.

EXAMPLE 3-3

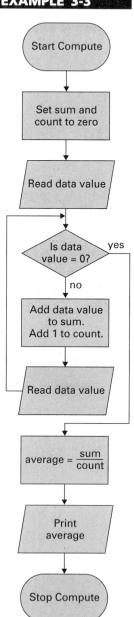

Average of a Set of Data Values

The following flowchart was developed for an algorithm to find the average of a set of data values. Convert the flowchart into a program.

SOLUTION 1

We have covered the FORTRAN statements for the If structures and for the While loop. Using these statements, we can translate each step in the flowchart into FORTRAN.

```
*------------------------------------------------------------------*
      PROGRAM COMPUT
*
*  This program computes and prints the average
*  of a set of experimental data values.
*
      INTEGER COUNT
      REAL SUM, X, AVERG
*
      SUM = 0.0
      COUNT = 0
*
      READ*, X
    1 IF (X.NE.0.0) THEN
         SUM = SUM + X
         COUNT = COUNT + 1
         READ*, X
         GO TO 1
      END IF
*
      AVERG = SUM/REAL(COUNT)
*
      PRINT 5, AVERG
    5 FORMAT (1X,'THE AVERAGE IS ',F5.2)
*
      END
*------------------------------------------------------------------*
```

SOLUTION 2

We now present a second solution that uses a logical variable DONE to specify when we have reached the end of the data. In this solution, we initialize the logical variable to the value .FALSE.; when we find the value that indicates the end of the data, we change the value of the logical variable to .TRUE. .

```
*-------------------------------------------------------------------*
      PROGRAM COMPUT
*
*  This program computes the average of
*  a set of experimental data values.
*
      INTEGER COUNT
      REAL SUM, X, AVERG
      LOGICAL DONE
*
      SUM = 0.0
      COUNT = 0
      DONE = .FALSE.
*
    1 IF (.NOT.DONE) THEN
         READ*, X
         IF (X.EQ.0.0) THEN
            DONE = .TRUE.
         ELSE
            SUM = SUM + X
            COUNT = COUNT + 1
         END IF
         GO TO 1
      END IF
*
      AVERG = SUM/REAL(COUNT)
*
      PRINT 5, AVERG
    5 FORMAT (1X,'THE AVERAGE IS ',F5.2)
*
      END
*-------------------------------------------------------------------*
```

. . . .

3-5 Application   ## ROCKET TRAJECTORY

Aerospace Engineering

A small rocket is being designed to make wind shear measurements in the vicinity of thunderstorms. Before testing begins, the designers are developing a simulation of the rocket's trajectory. They have derived the following equation, which they believe will predict the performance of their test rocket, where t is the elapsed time in seconds:

$$\text{height} = 60 + 2.13t^2 - 0.0013t^4 + 0.000034t^{4.751}$$

The equation gives the height above ground level at time t. The first term (60) is the height in feet above ground level of the nose of the rocket. To check the predicted performance, the rocket is "flown" on a computer, using the preceding equation.

Develop an algorithm and use it to write a complete program to cover a maximum flight of 100 seconds. Increments in time are to be 2.0 seconds from launch through the ascending and descending portions of the trajectory until the rocket descends to within 50 feet of ground level. Below 50 feet the time increments are to be 0.05 seconds. If the rocket impacts prior to 100 seconds, the program is to stop immediately after impact. The output is to be a table of corresponding time and height values.

As shown in the following diagram, several possible events could occur as we simulate the flight. The height above the ground should increase for a period and then decrease until the rocket impacts. We can test for impact by testing the height for a value equal to or less than zero. It is also possible that the rocket will still be airborne after 100 seconds of flight time. Therefore, we must also test for this condition and stop the program if the value of time becomes greater than 100. In addition, we need to observe the height above ground. As the rocket approaches the ground, we want to monitor its progress more frequently; we will need to reduce our time increment from 2.0 seconds to 0.05 seconds.

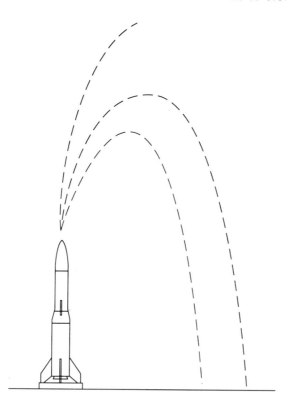

1. Problem Statement

Print a table of time and height values for a rocket trajectory. Start time at zero and increment it by 2.0 seconds until the height is less than 50 feet, then increment time by 0.05 seconds. Stop the program if the rocket impacts or if the total time exceeds 100 seconds.

2. Input/Output Description

Input—none

Output—a table of time and height values

3. Hand Example

Using a calculator, you compute the first three entries in our table as shown:

Time (s)	Height (ft)
0.0000	60.0000
2.0000	68.5001
4.0000	93.7719

We are now ready to develop the algorithm so that the computer can compute the rest of the table for us.

4. Algorithm Development

Decomposition

Set time to zero.
Compute and print times and heights.

We do the refinement for this problem in two steps:

Initial Pseudocode

Rocket1: time ← 0
  While above ground and time ≤ 100 do
    Compute height
    Print time, height
    Increment time

In our refinement we replace "Increment time" with the steps that take into account our height above ground. We also replace the condition "above ground" with a specific condition based on our height above ground.

Final Pseudocode

Rocket1: time ← 0
  height ← 60
  While height > 0 and time ≤ 100 do
    Compute height
    Print time, height
    If height < 50 then
      time ← time + 0.05
    Else
      time ← time + 2.0

Notice that we initialized the height variable to 60.0 before entering the While loop. Why? Could we have initialized it after the condition was tested? We must also set the height to a value greater than zero, or the While loop would never be executed.

FORTRAN Program 1

```
*------------------------------------------------------------------*
      PROGRAM RCKET1
*
*  This program simulates a rocket flight for up to 100 seconds.
*  A table contains height values at increments of 2 seconds
*  until the rocket is within 50 feet of the ground. Within 50
*  feet of the ground, the height values are computed every
*  0.05 seconds.
*
      REAL TIME, HEIGHT
*
      TIME = 0.0
      HEIGHT = 60.0
*
      PRINT 5
    5 FORMAT (1X,'TIME (S)     HEIGHT (FT)')
      PRINT*
*
   10 IF (HEIGHT.GT.0.0.AND.TIME.LE.100.0) THEN
         HEIGHT = 60.0 + 2.13*TIME**2 - 0.0013*TIME**4
     +             + 0.000034*TIME**4.751
         PRINT 15, TIME, HEIGHT
   15    FORMAT (1X,F7.4,8X,F9.4)
         IF (HEIGHT.LT.50.0) THEN
            TIME = TIME + 0.05
         ELSE
            TIME = TIME + 2.0
         END IF
         GO TO 10
      END IF
*
      END
*------------------------------------------------------------------*
```

As our programs become longer, we use comment lines with only an asterisk in column 1 to separate groups of statements that have a common function. In the preceding program, the blank comment lines separate steps that initialize variables and print the heading from the While loop.

5. Testing

The first few lines and the last few lines of output are shown. The values are in agreement with the hand example.

```
TIME (S)        HEIGHT (FT)
  0.0000            60.0000
  2.0000            68.5001
  4.0000            93.7719
  6.0000           135.1644
  8.0000           191.6590
     .
     .
     .
 54.0000           999.1536
 56.0000           827.4205
 58.0000           633.2993
 60.0000           418.3975
 62.0000           184.8088
 64.0000           -64.8608
```

Can you think of ways to test different parts of the algorithm? We now know that the rocket impacts before 100 seconds of flight time; we could change the cutoff time to 50 seconds to see if this exit from the While loop were working correctly. How could you modify the program to check the change in the increment of the time variable from 2.0 seconds to 0.05 seconds?

3-6 DO LOOP

In Section 3-4 we used the IF statement to build While loops. A special form of the While loop is the counting loop, or *iterative loop*. Implementing a counting loop generally involves initializing a counter before entering the loop, modifying the counter within the loop, and exiting the loop when the counter reaches a specified value. Counting loops are executed a specified number of times. The three steps (initialize, modify, and test) can be incorporated in a While loop, as we have already seen, but they still require three different statements. A special statement, the DO statement, combines all three steps into one. Using the DO statement to construct a loop results in a construction called a *DO loop*.

The general form of the DO statement is

> DO *k index* = *initial, limit, increment*

The constant *k* is the number of the statement that represents the end of the loop; *index* is a variable used as the loop counter; initial represents the *initial value* given to the loop counter; *limit* represents the value used to determine when the DO loop has been completed; and *increment* represents the value to be added to the loop counter each time the loop is executed.

The values of initial, limit, and increment are the *parameters* of the DO loop. If the increment is omitted, an increment of 1 is assumed. When the value of the index is greater than the limit, control is passed to the statement following the end of the loop. The end of the loop is usually indicated by the

CONTINUE statement, whose general form is

```
k CONTINUE
```

where *k* is the statement number referenced by the corresponding DO statement. Before listing the rules for using a DO loop, let's look at a simple example.

EXAMPLE 3-4   ## Integer Sum

The sum of the integers 1 through 50 is represented mathematically as

$$\sum_{i=1}^{50} i = 1 + 2 + \cdots + 49 + 50$$

Obviously, we do not want to write one long assignment statement to compute this sum. A better solution is to build a loop that executes 50 times and adds a number to the sum each time, as shown in the following solutions.

```
    While Loop Solution              DO Loop Solution
       SUM = 0                          SUM = 0
       NUMBER = 1                       DO 10 NUMBER=1,50
    10 IF (NUMBER.LE.50) THEN              SUM = SUM + NUMBER
          SUM = SUM + NUMBER        10 CONTINUE
          NUMBER = NUMBER + 1
          GO TO 10
       END IF
```

The DO statement identifies statement 10 as the end of the loop. The index NUMBER is initialized to 1. The loop is repeated until the value of NUMBER is greater than 50. Because the third parameter is omitted, the index NUMBER is incremented automatically by 1 at the end of each loop. Comparing the DO loop solution with the While loop solution, we see that the DO loop solution is shorter but that both compute the same value for SUM.

· ·

Structure of a DO Loop

We have seen a DO loop in a simple example. The following is a summary of the rules related to its structure.

1. The index of the DO loop must be a variable, but it may be either real or integer.
2. The parameters of the DO loop may be constants, variables, or expressions and can also be real or integer types.
3. The increment can be either positive or negative, but it cannot be zero.
4. A DO loop may end on any executable statement that is not a transfer, an IF statement, or another DO statement. The CONTINUE statement is an

executable statement that was designed expressly for closing a DO loop; although other statements may be used, we strongly encourage the consistent use of the CONTINUE statement to indicate the end of the loop.

5. Use the following pseudocode for a counting loop:

> For index = initial to limit by step do
>   set of statements

The "by step" clause is omitted if the increment is 1. The pseudocode for the loop in Example 3-4 is

> For number = 1 to 50 do
>   sum ← sum + number

6. The flowchart symbol for a DO loop is

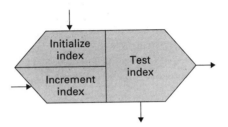

Note that the symbol is divided into three parts, each corresponding to the three steps in building an iterative loop. The flowchart for the iterative loop in Example 3–4 is

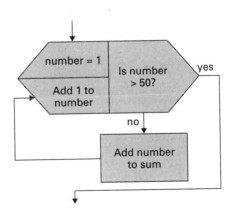

These rules define the structure of the DO loop but do not define the steps in its execution, which is covered next.

Execution of a DO Loop

In the following list we present a complete set of rules related to DO loop execution. We suggest that you read through them once and then go through the set of examples that follow the rules. Later, reread these rules. Do not worry about memorizing them; after you have gone through the set of examples, most of the rules will seem logical.

1. The test for completion is done at the beginning of the loop, as in a While loop. If the initial value of the index is greater than the limit and the increment is positive, the loop will not be executed. For instance, the statement

$$\text{DO 10 I=5,2}$$

sets up a loop that ends at statement 10. The initial value of the index I is 5, which is greater than the limit 2; therefore, the statements within the loop will be skipped and control passes to the statement following statement 10.

2. The value of the index should not be modified by other statements during the execution of the loop.

3. After the loop begins execution, changing the values of the parameters will have no effect on the loop.

4. If the increment is negative, the exit from the loop will occur when the value of the index is less than the limit.

5. Although it is not recommended, you may branch out of a DO loop before it is completed. The value of the index will be the value just before the branch. (If you need to exit the DO loop before it is completed, you should restructure the loop as a While loop to maintain a structured program.)

6. Upon completion of the DO loop, the index contains the last value that exceeded the limit.

7. Always enter a DO loop through the DO statement so that it will be initiated properly.

8. It is invalid to use a GO TO statement to transfer from outside a DO loop to inside the DO loop.

9. The number of times that a DO loop will be executed can be computed as

$$\left[\frac{\text{limit} - \text{initial}}{\text{increment}} \right] + 1$$

The brackets around the fraction represent the greatest integer value; that is, we drop any fractional portion (truncate) of the quotient. If this value is negative, the loop is not executed. If we had the following DO statement

$$\text{DO 35 K=5,83,4}$$

the corresponding DO loop would be executed the following number of times:

$$\left[\frac{83 - 5}{4} \right] + 1 = \left[\frac{78}{4} \right] + 1 = 20$$

The value of the index K would be 5, then 9, then 13, and so on until the final value of 81. The loop would not be executed with the value 85 because it is greater than the limit, 83.

The next set of examples illustrates both the structure and the execution of the DO loop.

EXAMPLE 3-5

Polynomial Model

Polynomials are often used to model data and experimental results. Assume that the polynomial $3t^2 + 4.5$ models the results of an experiment where t represents time in seconds. Write a program to evaluate this polynomial for time beginning at zero seconds and ending at 5 seconds in increments of 0.5 seconds.

SOLUTION

Step 1 is to state the problem clearly: Print a report to evaluate the polynomial $3t^2 + 4.5$ for a period from 0 seconds through 5 seconds in increments of 0.5 seconds.

Step 2 is to describe the input and output:

> Input — none
> Output — report containing polynomial values

Step 3 is to work a simple example. Therefore, the first few lines of output should be

Polynomial Model

Time	Polynomial
(S)	
0.0	4.50
0.5	5.25
1.0	7.50

Step 4 is to develop an algorithm, beginning with the decomposition.

Decomposition

> Print headings.
> Print report.

Pseudocode

```
Poly1: Print headings
       For k = 0 to 50 by 5 do
           time ← 0.1 · k
           poly ← 3 time² + 4.5
           Print time, poly
```

FORTRAN Program

```
*-----------------------------------------------------------------*
      PROGRAM POLY1
*
*  This program prints a table of values for a polynomial
*  starting with time equal to 0 seconds through 5 seconds
*  in increments of 0.5 seconds.
*
```

```
      INTEGER K
      REAL TIME, POLY
*
      PRINT*, 'POLYNOMIAL MODEL'
      PRINT*
      PRINT*, 'TIME    POLYNOMIAL'
      PRINT*, '(S)'
*
      DO 15 K=0,50,5
         TIME = 0.1*REAL(K)
         POLY = 3.0*TIME**2 + 4.5
         PRINT 10, TIME, POLY
   10    FORMAT (1X,F4.1,5X,F5.2)
   15 CONTINUE
*
      END
*-----------------------------------------------------------------*
```

Note that the value of TIME is computed from an integer that varies from 0 to 50 in steps of 5. Another possibility would have been to use the following DO statement:

```
                 DO 15 TIME=0.0,5.0,0.5
```

However, it is wise to avoid DO statements such as this one with real parameters. These statements do not always execute exactly the way we expect because of truncation within the computer. For example, suppose that the value for 0.5 is stored as a value slightly less than 0.5 in our computer system. Each time we add 0.5 to the index, we are adding less than we intend, and the values are not those we intend. Also, in this case the loop would be executed one more time than we intended because the value of TIME would not exactly equal 5.0 during the execution of the loop.

   Step 5 is to test the program. The output from this program is

```
POLYNOMIAL MODEL

TIME    POLYNOMIAL
(S)
0.0        4.50
0.5        5.25
1.0        7.50
1.5       11.25
2.0       16.50
2.5       23.25
3.0       31.50
3.5       41.25
4.0       52.50
4.5       65.25
5.0       79.50
```

Try It Try this self-test to check your memory of some key points from Sec⸱
If you have any problems with the exercises, you should reread this
The solutions are given at the end of this module.

In problems 1–6 determine the number of times that the statements in the
DO loop will be executed. Assume that the index is an integer variable.

1. `DO 10 NUM=5,14`   2. `DO 10 COUNT=-4,4`

3. `DO 10 K=15,3,-1`   4. `DO 10 TIME=-5,15,3`

5. `DO 10 TIME=50,250,25`   6. `DO 10 INDEX=72,432,4`

For problems 7–11, give the value in COUNT after each of the following
loops is executed. Assume that COUNT is an integer variable initialized to
zero before each problem.

```
  7.    DO 5 I=1,8
            COUNT = COUNT + 1
       5 CONTINUE
```
```
  8.    DO 5 K=1,5
            COUNT = COUNT + K
       5 CONTINUE
```
```
  9.    DO 5 INDEX=0,7
            COUNT = COUNT - 2
       5 CONTINUE
```
```
 10.    DO 5 NUM=8,0,-1
            COUNT = COUNT + 2
       5 CONTINUE
```
```
 11.    DO 5 M=5,5
            COUNT = COUNT + (-1)**M
       5 CONTINUE
```

3-7 Application   TIMBER REGROWTH

Environmental Engineering

A problem in timber management is to determine how much of an area to
leave uncut so that the harvested area is reforested in a certain period of
time. It is assumed that reforestation takes place at a known rate per year,
depending on climate and soil conditions. A reforestation equation expresses this growth as a function of the amount of timber standing and
the reforestation rate. For example, if 100 acres are left standing after
harvesting and the reforestation rate is 0.05, then $100 + 0.05 \times 100$, or
105 acres, are forested at the end of the first year. At the end of the second
year, the number of acres forested is $105 + 0.05 \times 105$, or 110.25 acres.

Write a program to read the identification number of an area, the total
number of acres in the area, the number of acres that are uncut, and the
reforestation rate. Print a report that tabulates for 20 years the number of
acres reforested and the total number of acres forested at the end of each
year.

1. Problem Statement

Compute the number of acres forested at the end of each year for 20 years
for a given area.

2. Input/Output Description

Input — the identification number for the area of land, the total acres, the number of acres with trees, and the reforestation rate.

Output — a table with a row of data for each of 20 years. Each row of information contains the number of acres reforested during that year and the total number of acres forested at the end of the year.

3. Hand Example

Assume that there are 14,000 acres total with 2500 acres uncut. If the reforestation rate is 0.02, we can compute a few entries as shown:

Year 1 $2500 \times 0.02 = 50$ acres of new growth
               original 2500 acres + 50 new acres = 2550 acres forested

Year 2 $2550 \times 0.02 = 51$ acres of new growth
               original 2550 acres + 51 new acres = 2601 acres forested

Year 3 $2601 \times 0.02 = 52.02$ acres of new growth
               original 2601 acres + 52.02 new acres = 2653.02 acres forested

4. Algorithm Development

The overall structure is a counting loop that is executed 20 times, once for each year. Inside the loop we need to compute the number of acres reforested during that year and add that number to the acres forested at the beginning of the year; this will compute the total number of acres forested at the end of the year. The output statement should be inside the loop, because we want to print the number of acres forested at the end of each year.

Decomposition

Read initial information.
Print headings.
Print report.

Initial Pseudocode

Timber: Read initial information
         Print headings
         For year = 1 to 20 do
                 Compute reforested amount
                 Add reforested amount to uncut amount
                 Print reforested amount, uncut amount

Clearly, an error condition exists if the uncut area exceeds the total area. We will test for this condition and exit after printing an error message if it occurs. It is possible to imagine soil conditions that would result in a zero or negative reforestation rate, so we will not perform any error checking on the rate. However, all the values read will be printed, or echoed, so that the user can recognize an error in an input value. We now add these refinements to our initial pseudocode.

Final Pseudocode

Timber: Read identification, total, uncut, rate
Print identification, total, uncut, rate
If uncut > total then
    Print error message
Else
    Print headings
    For year = 1 to 20 do
        refor ← uncut × rate
        uncut ← uncut + refor
        Print year, refor, uncut

FORTRAN Program

```
*--------------------------------------------------------------*
      PROGRAM TIMBER
*
*  This program computes a reforestation summary
*  for an area that has not been completely harvested.
*
      INTEGER ID, YEAR
      REAL TOTAL, UNCUT, RATE, REFOR
*
      PRINT*, 'ENTER LAND IDENTIFICATION (INTEGER)'
      READ*, ID
      PRINT*, 'ENTER TOTAL NUMBER OF ACRES'
      READ*, TOTAL
      PRINT*, 'ENTER NUMBER OF ACRES UNCUT'
      READ*, UNCUT
      PRINT* 'ENTER REFORESTATION RATE'
      READ*, RATE
*
      IF (UNCUT.GT.TOTAL) THEN
         PRINT*, 'UNCUT AREA LARGER THAN ENTIRE AREA'
      ELSE
         PRINT 5, ID, TOTAL, UNCUT, RATE
    5    FORMAT (/'REFORESTATION SUMMARY'//
     +           1X,'IDENTIFICATION NUMBER ',I5/
     +           1X,'TOTAL ACRES = ',F10.2/
     +           1X,'UNCUT ACRES = ',F10.2/
     +           1X,'REFORESTATION RATE = ',F5.3//
     +           1X,'YEAR  REFORESTED  TOTAL REFORESTED')
         DO 15 YEAR = 1,20
            REFOR = UNCUT*RATE
            UNCUT = UNCUT + REFOR
            PRINT 10, YEAR, REFOR, UNCUT
   10       FORMAT (1X,I3,F11.3,F17.3)
   15    CONTINUE
      END IF
```

```
*
      END
*_____*
```

5. Testing

Using the test data from the hand example, a typical interaction is

```
ENTER LAND IDENTIFICATION (INTEGER)
25563
ENTER TOTAL NUMBER OF ACRES
14000.0
ENTER NUMBER OF ACRES UNCUT
2500.0
ENTER REFORESTATION RATE
0.02

REFORESTATION SUMMARY

IDENTIFICATION NUMBER 25563
TOTAL ACRES =   14000.00
UNCUT ACRES =    2500.00
REFORESTATION RATE = 0.020

YEAR  REFORESTED   TOTAL REFORESTED
  1     50.000        2550.000
  2     51.000        2601.000
  3     52.020        2653.020
  4     53.060        2706.080
  5     54.122        2760.202
  6     55.204        2815.406
  7     56.308        2871.714
  8     57.434        2929.148
  9     58.583        2987.731
 10     59.755        3047.486
 11     60.950        3108.436
 12     62.169        3170.604
 13     63.412        3234.017
 14     64.680        3298.697
 15     65.974        3364.671
 16     67.293        3431.964
 17     68.639        3500.604
 18     70.012        3570.615
 19     71.412        3642.028
 20     72.841        3714.868
```

The numbers match the ones we computed by hand. Try an example to test the error condition by using an uncut area larger than the total area. What happens if the reforestation rate is 0.00 or − 0.02? Should there be an upper limit on the reforestation rate? This information is not given in the original problem, so we probably should not set one arbitrarily. What happens if you enter 14,000.0 instead of 14000.0? It might be a good idea to remind the program user not to use commas in numbers. How would you do this?

3-8   NESTED DO LOOPS

DO loops can be nested within other DO loops, just as we used IF structures within other IF structures. The following rules apply to writing and executing *nested DO loops:*

1. A nested DO loop cannot use the same index as a loop that contains it.
2. A nested DO loop must be completely within the outer DO loop. For example, if the DO statement is within another DO loop, the CONTINUE statement for the nested loop must also be within the outer loop.
3. DO loops that are independent of each other may use the same index, even if they are all contained within another DO loop.
4. When one loop is nested within another, the inside loop is completely executed with each pass through the outer loop.
5. Although nested DO loops can end on the same CONTINUE statement, we recommend that you end each loop with a separate CONTINUE so that the indenting can be consistent.

The following are a set of valid loops and a set of invalid loops:

Valid Loops

```
      DO 15 I=1,5              DO 15 I=1,5
         DO 10 J=1,8              DO 10 K=1,8
            DO 5 K=2,10,2            ...
               ...          10    CONTINUE
    5       CONTINUE             DO 5 K=2,10,2
   10    CONTINUE                   ...
   15 CONTINUE             5    CONTINUE
                          15 CONTINUE
```

Invalid Loops

```
      DO 15 I=1,5              DO 20 J=1,5
         DO 10 J=1,8              DO 10 J=1,8
            DO 5 K=2,10,2            ...
               ...          10    CONTINUE
   10    CONTINUE                DO 15 K=2,10,2
         ...                        ...
    5       CONTINUE        15    CONTINUE
   15 CONTINUE             20 CONTINUE
    (overlapping  loops)       (same  index  for
                                 dependent  loops)
```

To illustrate the execution of a program with nested loops, consider the following:

```
*-------------------------------------------------------------------*
      PROGRAM NEST
*
*  This program prints the indices in nested DO loops.
```

```
*
      INTEGER I, J
*
      PRINT*, '  I      J'
      PRINT*
      DO 20 I=1,5
         DO 10 J=3,1,-1
            PRINT 5, I, J
    5       FORMAT (1X,I3,3X,I3)
   10    CONTINUE
         PRINT*, '  END OF PASS'
   20 CONTINUE
*
      END
*---------------------------------------------------------------*
```

The output is the following:

```
I       J

1       3
1       2
1       1
END OF PASS
2       3
2       2
2       1
END OF PASS
3       3
3       2
3       1
END OF PASS
4       3
4       2
4       1
END OF PASS
5       3
5       2
5       1
END OF PASS
```

The first time through the outer loop, I is initialized to the value 1. We thus begin executing the inner loop: The variable J is initialized to the value 3. After executing the PRINT 5 statement, we reach the end of the inner loop and J is decremented by 1 to the value 2. Because 2 is still larger than the final value of 1, we repeat the loop. J is decremented by 1 to the value 1, and we repeat the loop again. When J is decremented again, it is less than the final value of 1, so we have completed the inner loop and the message 'END OF PASS' is printed. I is incremented to 2, and we begin the inner loop again. This process is repeated until I is greater than 5.

EXAMPLE 3-6   ## Experimental Sums

Write a complete program to read 20 data values. Compute the sum of the first 5 values, the next 5 values, and so on. Print the 4 sums. Assume that the values are real.

SOLUTION

Step 1 is to state the problem clearly: Read 20 data values and compute the sum of the first 5 values, the second 5 values, and so on.

Step 2 is to describe the input and output:

> Input—20 data values
> Output—sum of the first 5 values, sum of the second set of
> 5 values, and so on for all 20 data values

Step 3 is to work a simple example by hand. Assume we have 6 values and we want to sum them in groups of 2. Then we have 3 sums to compute, and for each sum we add 2 values.

$$\text{values} - \quad 4, 6, 1, -2, 7, -2$$

The corresponding sums are $10, -1, 5$

Step 4 is to develop an algorithm. Although we will read a total of 20 data values, we need only 5 values at a time; thus, we need an outer loop to read four sets of data. Each set of data is 5 values; thus, the inner loop reads the 5 values. It is important to initialize to zero the variable being used to store the sum before the inner loop is begun. We then add 5 values, print the sum, and set the sum back to zero before we read the next values.

The decomposition is a single step in this problem; all the other steps are performed inside the overall loop.

Decomposition

> Read data, and compute and print sums.

The refinement in pseudocode illustrates the structure of this algorithm, which is a loop within a loop.

Pseudocode

> Sums: For $i = 1$ to 4 do
>   sum $\leftarrow$ 0
>   For $j = 1$ to 5 do
>   Read value
>   sum $\leftarrow$ sum + value
>   Print sum

We can now translate the pseudocode into FORTRAN.

FORTRAN Program

```
*-------------------------------------------------------------------*
      PROGRAM SUMS
*
*  This program reads 20 values and prints
*  the sum of each group of 5 values.
*
```

```
         INTEGER I, J
         REAL SUM, VALUE
  *
         DO 15 I=1,4
            SUM=0.0
  *
            DO 5 J=1,5
               READ*, VALUE
               SUM = SUM + VALUE
  5         CONTINUE
  *
            PRINT 10, I, SUM
  10        FORMAT(1X,'SUM ',I1,' = ',F6.2)
  15    CONTINUE
  *
         END
  *_____*
```

Step 5 is to test the program. Here is a sample of the type of output that would be printed from this program after the 20 values have been entered.

```
SUM 1 =   23.44
SUM 2 =   10.23
SUM 3 =   -5.69
SUM 4 =    1.01
```

. . . .

EXAMPLE 3-7   ## Factorial Computation

Write a complete program to compute the factorial of an integer. A few factorials and their corresponding values are shown (an exclamation point following a number symbolizes a factorial):

$$0! = 1$$
$$1! = 1$$
$$2! = 2 \times 1$$
$$3! = 3 \times 2 \times 1$$

Compute and print the factorial for four values read from the terminal.

SOLUTION

Step 1 is to state the problem clearly: Compute the factorial of values read from the terminal.

Step 2 is to describe the input and output:

> Input—four data values
> Output—factorial of each of the four data values

Step 3 is to work a simple example by hand:

$$5! = 5 \times 4 \times 3 \times 2 \times 1 = 120$$

Step 4 is to develop an algorithm. Because the factorial of a negative number is not defined, we should include an error check in the algorithm for this condition. In computing a factorial, we use a counting loop to perform the successive multiplications. Because the overall structure of this problem solution is a loop, the decomposition is again a single step.

Decomposition

> Read values, and compute and print factorials.

Pseudocode

```
Fact: For i = 1 to 4 do
          Read n
          If n < 0 then
                  Print error message
          Else
                  nfact ← 1
                  If n > 1 then
                          For k = 1 to n do
                                  nfact ← nfact × k
                  Print n, nfact
```

FORTRAN Program

```
*------------------------------------------------------------------*
      PROGRAM FACT
*
*   This program computes the factorial
*   of four values read from the terminal.
*
      INTEGER I, N, NFACT, K
*
      DO 20 I=1,4
          PRINT*, 'ENTER N'
          READ*, N
*
          IF (N.LT.0) THEN
*
              PRINT 5, N
    5         FORMAT (1X,'INVALID N = ',I7)
*
          ELSE
*
              NFACT = 1
              IF (N.GT.1) THEN
                  DO 10 K=1,N
                      NFACT = NFACT*K
   10             CONTINUE
              END IF
              PRINT 15, N, NFACT
   15         FORMAT (1X,I4,'! = ',I8)
*
          END IF
   20 CONTINUE
```

```
*
        END
*-------------------------------------------------------------------*
```

Step 5 is to test the program. The output from a sample run of this program is

```
ENTER N
3
    3! =         6
ENTER N
-2
INVALID N =       -2
ENTER N
11
  11! = 39916800
ENTER N
0
    0! =         1
```

. . . .

Try It

!

Try this self-test to check your memory of some key points from Section 3-8. If you have any problems with the exercises, you should reread this section. The solutions are included at the end of this module.

For problems 1–3, give the value in COUNT after each of the following loops is executed. Assume that COUNT is initialized to zero before starting each problem.

```
1.    DO 10 IN=5,15
         DO 5 K=2,0,-1
            COUNT = 100 + 1
   5     CONTINUE
  10 CONTINUE
2.    DO 10 K=0,5
         DO 5 J=5,-5,-2
            COUNT = COUNT + 1
   5     CONTINUE
  10 CONTINUE
3.    DO 10 I=5,14,2
         DO 5 K=4,0
            COUNT = COUNT + 1
   5     CONTINUE
         COUNT = COUNT + 1
  10 CONTINUE
```

SUMMARY

This chapter greatly expanded the types of problems we can solve in FOR-TRAN by using IF statements: We can now control the order in which state-

ments are executed. An important property of these statements is
are entered only at the top of the structure and they have only one
only one entrance and one exit, this type of flow promotes the v
simpler programs. We also learned to use both While loops and DC   ᵖᵘ ᵗᵒ
implement repetition steps. Most of the programs in the rest of the module
will use either one or the other of these loop structures.

Key Words

branch
compound logical expression
condition
control structure
counting loop
DO loop
flowchart
hierarchy
IF structure
increment value
index
initial value
iterative loop
limit value

logical expression
logical operator
logical variable
loop
nested DO loop
parameter
pseudocode
relational operator
repetition
selection
sequence
structured algorithm
structured program
While loop

Problems

This problem set begins with modifications to programs given earlier in this
chapter. Give the decomposition, refined pseudocode or flowchart, and
FORTRAN program for each problem.

Problem 1 modifies the light pipe program PIPE, given in Section 3-3.

1. Modify the light pipe program so that it reads the refraction index for the
   rod material. Then, for an incidence angle of 45°, give the range of refraction indices for materials that will create light pipes.

Problem 2 modifies the trajectory program RCKET1, given in Section 3-5.

2. Modify the rocket trajectory program so that it reads two values, INCR1
   and INCR2. Start time at zero seconds and increment it by INCR1 seconds until the distance is less than 50 feet; then increment time by
   INCR2 seconds.

Problems 3–4 modify the regrowth program TIMBER, given in Section 3-7.

3. Modify the timber regrowth program so that a value of M is read from the
   terminal, where M represents the number of years that should be between lines in the output table.

4. Modify the timber regrowth program so that instead of printing data for 20 years, it prints yearly information until at least 10 percent of the cut area has been reforested.

For problems 5–7, write complete FORTRAN programs to print tables showing the values of the input variables and the function shown using DO loops to control the loops.

5. Print a table of values for K where

$$K = |J - 3J + \sqrt{J}|$$

for values of $J = 0, 1, 2, \ldots ,N$, where N is read from the terminal.

6. Print a table of values for Y where

$$Y = \cos X$$

for values of $X = 0.0, 10.0, \ldots , 350.0$ where the units of X are in degrees.

7. Print a table of values for F where

$$F = XY - 1$$

for values of $X = 1, 2, \ldots , 9$ and values of $Y = 0.5, 0.75, 1.0, \ldots , 2.5$.

Develop new programs in problems 8–12. Use the five-step design process.

8. Write a program to read three integers I, J, K. Then, determine if the integers are in ascending order ($I \leq J \leq K$), or if the integers are in descending order ($I \geq J \geq K$), or if they are in neither order. Print an appropriate message.

9. A set of 20 temperatures has been recorded during an experiment. We want to compute and print the average temperature, and we also want to determine if a new record low or record high temperature occurred during this particular experiment. Therefore, on one line we enter the previous low temperature and the previous high temperature. The remaining 20 temperatures are entered one per line. Read this information and print the average. If a new record low or high occurred, print an appropriate message.

10. Consider the trajectory of a stone hurled into the air with an initial velocity v at an angle to the horizontal of θ. Neglecting drag due to friction with the atmosphere, the following equations describe the stone's distance (d) from the initial spot and the height (h) of the stone at time t:

$$d = v \cdot t \cos \theta$$

$$h = v \cdot t \sin \theta - \frac{1}{2} g \cdot t^2$$

where g is the acceleration due to gravity (9.8 m/s^2 or 32 ft/s^2). Write a program to read the initial velocity and angle and then print a table of distance and height values. Use intervals of 0.25 seconds between values and stop when the height becomes negative.

11. A biologist, after discovering the omega bacterium, has spent 5 years determining its characteristics. She has found that the bacterium has a constant growth rate. If 10 cells are present with a growth factor of 0.1, the next generation will have $10 + 10(0.1) = 11$ cells. Write a program that will print a report with the following format:

```
                   OMEGA BACTERIA GROWTH

                   NUMBER OF CELLS   PETRI DISH        GROWTH
   CULTURE NUMBER   INITIALLY        DIAMETER (CM)     RATE

        XXXX          XXXX              XXX            XX.XX

   GENERATION  NUMBER OF CELLS   % AREA OF PETRI DISH COVERED
        1           XXXX.X                 XXX.XX
        .
        .
        .
        5
```

Ten cells occupy 1 square millimeter. Use the following input data for four cultures:

	NUMBER OF CELLS	PETRI DISH	GROWTH
CULTURE NUMBER	INITIALLY	DIAMETER (M)	RATE
1984	100	10 cm	0.50
1776	1300	5 cm	0.16
1812	600	15 cm	0.55
1056	700	8 cm	0.80

with title OMEGA BACTERIA GROWTH above.

12. The square of the sine function can be represented by the following:

$$\sin^2 x = x^2 - \frac{2^3 x^4}{4!} + \frac{2^5 x^6}{6!} - \cdots = \sum_{n=1}^{\infty} \frac{(-1)^{n+1} 2^{2n-1} x^{2n}}{(2n)!}$$

Write a program to evaluate this series for an input value, x, printing the results after $2, 4, 6, 8, \ldots, 14$ terms and comparing each sum to the true solution. Note that the term for $n = 1$ is x^2 and that all consecutive terms can be obtained by multiplying the previous term by

$$\frac{-(2x)^2}{2n(2n - 1)}$$

The output should have the following form:

COMPARISON OF VALUES OF SINE SQUARED

NUMBER OF TERMS	SERIES SUMMATION	INTRINSIC FUNCTION	ABSOLUTE DIFFERENCE
2	XX.XXXX	XX.XXXX	XX.XXXX
4	XX.XXXX	XX.XXXX	XX.XXXX
.			
.			
.			
14			

4 Engineering and Scientific Data Files

Critical Path Analysis | Critical path analysis is a technique used to determine the time schedule for large projects, ranging from the construction of a large building to the construction of the new 777 jumbo jet. This information is important in the planning stages before a project is begun, and it is also useful to evaluate the progress of a project that is partially completed. One method for this analysis starts by breaking a project into sequential events, and then breaking each event into various tasks. Although one event must be completed before the next one is started, various tasks within an event can occur simultaneously. The time it takes to complete an event, therefore, depends on the number of days required to finish its longest task. Similarly, the total time it takes to finish a project is the sum of the times it takes to finish the individual events.

INTRODUCTION ngineers and scientists must frequently analyze large amounts of data. It is not reasonable to enter all this data into the computer by hand each time it is needed. We can enter the data once into a data file; then each time the data is needed, we can read it from the data file. In addition to printing data for a report, we can write it into a data file; other programs can then easily access the data. This chapter focuses on reading information from existing data files.

4-1 I/O STATEMENTS

In the programs in previous chapters, we used READ statements when we needed information entered during the execution of the program. For small amounts of data, this interaction through the keyboard is satisfactory. However, for large amounts of data it is not feasible to enter the data every time we need to run the program. A *data file* is a file that contains data that can be read by a program. A data file can be built in two ways: It can be generated by another program, or it can be generated using a word processor or an editor. Each line of the data file is a *record*. A record in a FORTRAN 77 program file contains a computer language instruction, while a record in a data file contains data values. Once a file is generated, it can be listed or updated using an editor or a word processor. For now, our data files will contain only numbers; Chapter 8 covers how to work with character information.

To use data files with our programs, we must use some new statements and some extensions of old statements. Each of these statements references the filename that is assigned when a file is built. If we build a data file with a word processor or an editor, we assign the filename when we enter the data. If we build a data file with a program, we include a statement in the program that gives the file a name.

If a file is going to be used in a program, it must be opened before any processing can be done with the file. The OPEN statement gives the program several important pieces of information about the file, including the name of the file and whether the file is an input file or an output file. In addition, the OPEN statement connects the file to the program using a unit number that corresponds to the file. Then, within the program, whenever we want to refer to the file, we use the unit number. Since there may be several files used in the same program, the unit number gives us a method for uniquely specifying a file. The general form of the OPEN statement that we use in this chapter is the following:

> OPEN (UNIT=*integer expression*, FILE=*filename*, STATUS=*literal*)

The integer expression designates the unit number assigned to the file, the filename refers to the name given to the file when it was built, and the STATUS literal tells the computer whether the file is an input file or an output file. If the file is an input file, it is specified with STATUS='OLD'; if the file is an output file, it is specified with STATUS='NEW'.

The OPEN statement must precede any READ or WRITE statements that use the file. Also, the OPEN statement should be executed only once; therefore, it should not be inside a loop. Some systems require a REWIND statement after opening an input file to position the file at its beginning.

To read from a data file, we use an extension of the list-directed READ statement that has the following general form:

> READ *(unit number,*) variable list*

To write information to a data file, we use a new statement — a WRITE statement. Just like the PRINT statement, the WRITE statement can be used with list-directed output or formatted output. The list-directed WRITE statement has the following general form:

> WRITE *(unit number,*) expression list*

The formatted WRITE statement has the following general form:

> WRITE *(unit number, k) expression list*

where *k* is the statement number of the corresponding FORMAT statement. In all of these general forms, the unit number corresponds to the unit number assigned in the OPEN statement. The asterisk following the unit number specifies that we are using list-directed input or output.

Most computer systems have several input or output devices attached to them. Each device is assigned a unit number. For example, if a laser printer has been assigned unit number 8, then the following statement would write the values of X and Y using the laser printer:

> WRITE (8,*) X, Y

Many systems assign the standard input device (typically the terminal keyboard) to unit number 5 and the standard output device (typically the terminal screen) to unit number 6; these devices are used when your program executes READ* or PRINT* statements. Avoid using preassigned unit numbers with your data files; if your computer system assigns unit numbers 5 and 6 as previously defined, do not use them with your data files. Otherwise, errors could result if your program assigned a data file to unit 6 and the program then executed a PRINT* statement. The name of the file is determined when the file is built, but we may choose any unit number to refer to the file as long as the number is not preassigned by our computer system. Thus, in one program we could use unit 10 to refer to a file of experimental velocity measurements, and then in another program we could use unit 12 to refer to the same file.

When we finish executing a program that has used files, the files are automatically closed before the program terminates. Occasionally there are situations in which you would like to specifically close, or disconnect a file from your program. The FORTRAN statement to close a file has the following general form:

CLOSE (UNIT=*integer expression*)

We now summarize some important rules to remember when reading data from data files:

1. Each READ statement will start with a new line of data. If there are values left on the previous line that were not read, they will be skipped.
2. If a line does not contain enough values to match to the list of variables on the READ statement, another data line will automatically be read to acquire more values. Additional data lines will be read until values have been acquired for all the variables listed on the READ statement.
3. A READ statement does not have to read all the values on the data line; however, it does have to read all the values on the line prior to the values that you want. For example, if a file has five values per line and you are interested in the third and fourth values, you must read the first and second values to get to the third and fourth values, but you do not need to read the fifth value.

To use the correct READ statement, you must know how the values were entered in the data file. For example, assume that a data file contains two numbers per line, representing a time and a temperature measurement. Also assume that the first three lines of the file contain the following information:

line 1:	0.0	89.5
line 2:	0.1	90.3
line 3:	0.2	90.8

The following statement will correctly read a time and temperature value from the data file:

```
READ (10,*) TIME, TEMP
```

However, it would be incorrect to use the following two statements in place of the previous statement:

```
READ (10,*) TIME
READ (10,*) TEMP
```

The execution of these two statements reads two lines from the data file, not one. For example, if the two statements are the first READ statements to be executed using this file, the value of TIME will be 0.0 and the value of TEMP will be 0.1. Not only will we have the wrong values in the variables, but also there will not be an error message to let us know that something is wrong. This example illustrates the importance of carefully testing our programs with known data before using them with other data files.

Try It   Try this self-test to check your memory of some key points from Section 4-1. If you have any problems with the exercises, you should reread this section. The solutions are included at the end of this module.

The data file LAB1 contains the following information, which represents a time and temperature for each line:

line 1:	0.0	86.3
line 2:	0.5	93.5
line 3:	1.0	95.5
line 4:	1.5	97.8
line 5:	2.0	98.2
line 6:	2.5	99.1

Give the values of the variables after executing these statements. For each problem assume that the data file has been opened but no previous READ statements have been executed.

```
1. READ (10,*) TIME, TEMP
2. READ (10,*) TIME
   READ (10,*) TEMP
3. READ (10,*) TIME1, TEMP1, TIME2, TEMP2
4. READ (10,*) TIME1, TEMP1
   READ (10,*) TIME2, TEMP2
5. READ (10,*) TIME1
   READ (10,*) TEMP1
   READ (10,*) TIME2
   READ (10,*) TEMP2
6. READ (10,*) TIME1, TIME2
   READ (10,*) TEMP1, TEMP2
```

4-2   TECHNIQUES FOR READING DATA FILES

To read information from a data file, we must first know some information about the file. In addition to the name of the file, we must know what information is stored in the file. This information must be very specific, such as the number of values entered per line and the units of the values. In addition, we need to know if there is any special information in the file that will be useful in deciding how many records are in the file, or how to determine when we have read the last record. This information is important because if we execute a READ statement after we have already read the last record in the file, we will get an execution error and the program will quit. We can avoid an execution error by using the information that we have about the file to decide what kind of loop we should use in the program. For example, if we know that there are 200 records in the file, then we can use a DO loop that is executed 200 times, and each time through the loop we read a record and perform the desired computations with the information. Often, we do not know ahead of time how many records are in the data file, but we know that the last record contains special values so that our program can test for them. For example, if a file contained time and temperature measurements, the last record could contain values of -999 for both measurements to signal that it is the last record. Then, we could exit a While loop when the time and temperature values are -999. Finally, a third situation arises when we do not know ahead of time how many records are in a file, and there is not a special value at the end of the file. A special option in the READ statement allows us to handle this situation.

We now look at each of these three cases separately. For each case we use the same example so that we can compare the solutions.

Specified Number of Records

The first case we consider is one in which we know or we can ascertain the number of records in the data file. If we know the exact number of records, a DO loop can be used to process the data. Sometimes, we generate files with special information, such as the number of valid data records, in the first record in the data file. Then, each time we update the data file, we update the number in the first record. To process this file, we read the number at the beginning of the file using a variable, such as COUNT. Then we can execute a DO loop with COUNT as part of the DO statement. The following example illustrates reading a data file with a record count in the first record.

EXAMPLE 4-1   ## Average of Laboratory Data

Assume that we have a data file named RESULTS1 that contains information collected from a laboratory experiment. Each line of the data file, except the first line, contains two numbers: a time value in seconds and a temperature measurement in degrees Fahrenheit. Read the data and compute the average temperature value. Print the number of data values and the average value.

SOLUTION

Step 1 is to state the problem clearly: Compute and print the average of a set of temperature measurements.

Step 2 is to describe the input and output:

Input — time and temperature values in a data file named RESULTS1
Output — the average of the temperature values and a count of the temperature values

Step 3 is to work a simple example by hand. Assume that the data file contains the following information:

```
4
0.0        120
0.5        132
1.0        144
1.5        163
```

For this data, the total number of values is 4, and the average temperature is $(120 + 132 + 144 + 163)/4$, or $139.75°$ Fahrenheit.

Step 4 is to develop an algorithm, starting with the decomposition.

Decomposition

Read and compute a sum of temperature values.
Compute the average temperature.
Print the count and the average temperature.

Pseudocode

Soln1: sum ← 0
     Read count
     If count < 1 then
         Print error message
     Else
         For j = 1 to count do
              Read time, temperature
              sum ← sum + temperature

$$average \leftarrow \frac{sum}{count}$$

         Print count, average

FORTRAN Program

```
*------------------------------------------------------------------*
      PROGRAM SOLN1
*
* This program computes the average temperature using a set of
* times and temperatures stored in a data file.  The first line
* contains the number of actual data records that follow.
*
      INTEGER COUNT, J
      REAL SUM, TIME, TEMP, AVE
*
      OPEN (UNIT=10,FILE='RESULTS1',STATUS='OLD')
*
      READ (10,*) COUNT
      IF (COUNT.LT.1) THEN
*
         PRINT*, 'NO DATA ACCORDING TO RECORD COUNT'
*
      ELSE
*
         SUM = 0.0
         DO 20 J=1,COUNT
            READ (10,*) TIME, TEMP
            SUM = SUM + TEMP
   20    CONTINUE
         AVE = SUM/REAL(COUNT)
         PRINT 25, COUNT, AVE
   25    FORMAT (1X,'COUNT = ',I3,5X,'AVERAGE = ',F8.2,
     +           ' DEGREES F')
*
      END IF
```

```
*
      END
*------------------------------------------------------------------*
```

Step 5 is to test the program. If we test this program using the hand example, the following information is printed:

```
COUNT =    4     AVERAGE =    139.75 DEGREES F
```

. . . .

Trailer or Sentinel Signals

Special data values that are used to signal the last record of a file are called *trailer,* or *sentinel, signals.* When we generate the file, we add the special data values in the last record. Then, if we add or delete records from the file, we do not need to update a count; we only need to be sure that the last record still contains the special data values. When using a file with a trailer signal, we need to be careful that our program does not treat the information on the trailer line as regular data. For example, if we are counting the number of valid lines of data in the file, we want to be sure that we do not count the trailer line. Also, it is important to include the same number of data values on the trailer line as are on the rest of the lines in the data file. Since the trailer line is read with the statement that reads the other lines, it will try to read the line after the trailer line if there are not enough values on the trailer line; this will cause an execution error.

The following example illustrates reading a data file with a trailer record.

EXAMPLE 4-2 ## Average of Laboratory Data

Assume that we have a data file named RESULTS2 that contains information collected from a laboratory experiment. Each line of the data file contains two numbers: a time value in seconds and a temperature measurement in degrees Fahrenheit. However, the last line of the file is a trailer signal that contains -999.0 for the time value and temperature values. Read the data and compute the average value. Print the number of data values and the average value.

SOLUTION

This example resembles Example 4-1, except that we need to initialize the counter to zero and end the loop when we reach the trailer signal.

Pseudocode

```
Soln2: sum ← 0
       count ← 0
       Read time, temperature
       While time ≠ −999.0 then
             sum ← sum + temperature
             count ← count + 1
             Read time, temperature
```

If count = 0 then
   Print error message
Else

$$average \leftarrow \frac{sum}{count}$$

   Print count, average

FORTRAN Program

```
*-----------------------------------------------------------*
      PROGRAM SOLN2
*
*  This program computes the average temperature using a set of
*  times and temperatures stored in a data file.  The data file
*  contains a trailer line containing -999.0 for time and
*  temperature.
*
      INTEGER COUNT
      REAL TIME, TEMP, SUM, AVE
*
      OPEN (UNIT=10,FILE='RESULTS2',STATUS='OLD')
*
      SUM = 0.0
      COUNT = 0
*
      READ (10,*) TIME, TEMP
   5  IF (TIME.NE.-999.0) THEN
         SUM = SUM + TEMP
         COUNT = COUNT + 1
         READ (10,*) TIME, TEMP
         GO TO 5
      END IF
*
      IF (COUNT.EQ.0) THEN
         PRINT*, 'NO DATA VALUES IN THIS FILE'
      ELSE
         AVE = SUM/REAL(COUNT)
         PRINT 25, COUNT, AVE
   25    FORMAT (1X,'COUNT = ',I3,5X,'AVERAGE = ',F8.2,
     +           ' DEGREES F')
      END IF
*
      END
*-----------------------------------------------------------*
```

. . . .

END Option

If we do not know the number of data lines in a file, and if there is not a trailer
signal at the end, we must use a different technique. The following are the

steps, in pseudocode, that we want to perform:

> While there is more data do
> Read variables
> Process variables

To implement this in FORTRAN, we use an option available with the READ statement. This option tests for the end of the data and branches to a specified statement if the end is detected. A list-directed READ statement that uses this option has the following general form:

> READ *(unit number,*,*END=*n) variable list*

As long as there is data to read in the data file, this statement executes exactly like this statement:

> READ *(unit number,*) variable list*

However, if the last line of the data file has already been read and we execute the READ statement with the *END option,* then instead of getting an execution error, control is passed to the statement referenced in the END option. If the READ statement is executed a second time after the end of the data has been reached, an execution error will occur.

A READ statement with the END option is actually a special form of the While loop. When using it in a program, use the same indenting style as used in the While loop. A GO TO statement is also needed to complete the While loop, as shown:

```
    5 READ (10,*,END=15) TIME, TEMP
          SUM = SUM + TEMP
          COUNT = COUNT + 1
          GO TO 5
   15 PRINT*
```

This special implementation of the While loop should be used only when you do not know the number of data lines to be read and there is no trailer signal at the end of the file. If you have a trailer signal, test for that value to exit the loop instead of using the END option. The choice of the correct technique for reading data from a data file depends on the information in the data file. We can now solve the problem of computing an average of laboratory data for a file with no trailer signal and no way to determine the number of records in the file before running the program.

EXAMPLE 4-3 ## Average of Laboratory Data

Assume that we have a data file named RESULTS3 that contains information collected from a laboratory experiment. Each line of the data file contains two numbers — a time value in seconds and a temperature measurement in degrees Fahrenheit. There is no special information in the first record containing the number of lines in the file, and there is no trailer signal. Read the data and compute the average value. Print the number of data values and the average value.

SOLUTION

In this example, we use the END option to branch to the statement that performs the calculations.

Pseudocode

Soln3: sum ← 0
    count ← 0
    While more data do
        Read time, temperature
        sum ← sum + temperature
        count ← count + 1
    If count = 0 then
        Print error message
    Else

$$\text{average} \leftarrow \frac{\text{sum}}{\text{count}}$$

        Print count, average

FORTRAN Program

```
*------------------------------------------------------------------*
      PROGRAM SOLN3
*
*  This program computes the average temperature using a set of
*  times and temperatures stored in a data file.  The data file
*  contains only the time and temperature values.
*
      INTEGER COUNT
      REAL SUM, TIME, TEMP, AVE
*
      OPEN (UNIT=10,FILE='RESULTS3',STATUS='OLD')
*
      SUM = 0.0
      COUNT = 0
*
    5 READ (10,*,END=15) TIME, TEMP
         SUM = SUM + TEMP
         COUNT = COUNT + 1
         GO TO 5
*
   15 IF (COUNT.EQ.0) THEN
         PRINT* 'NO DATA VALUES IN THE FILE'
      ELSE
         AVE = SUM/REAL(COUNT)
         PRINT 25, COUNT, AVE
   25    FORMAT (1X,'COUNT = ',I3,5X,'AVERAGE = ',F8.2,
     +            ' DEGREES F')
      END IF
```

```
*
        END
*---------------------------------------------------------------*
```

. . . .

Try It

Try this self-test to check your memory of some key points from Section 4-2. If you have any problems with the exercises, you should reread this section. The solutions are included at the end of this module.

Assume that a data file contains time and altitude measurements from a rocket trajectory. Each line of the data file contains a corresponding time and altitude measurement. A trailer signal with a time equal to -99.0 is used to signal the end of the data file. Assume that the program contains the following statements:

```
          READ (10,*) TIME, ALT
        5 IF (TIME.NE.-99.0) THEN
              PRINT*, TIME, ALT
              READ (10,*) TIME, ALT
              GO TO 5
          END IF
```

For each of the following data lines, indicate if the line would work properly as the trailer line. If the line would not work properly, explain why it would not work.

1. -99.0 -99.0 2. -99.0 100.0

3. 100.0 -99.0 4. -99.0

5. -99.0 100.0 15.7

4-3 Application   **CRITICAL PATH ANALYSIS**

Manufacturing Engineering

A critical path analysis is a technique used to determine the time schedule for a project. This information is important in the planning stages before a project is begun, and it is also useful to evaluate the progress of a project that is partially completed. One method for this analysis starts by breaking a project into sequential events, then breaking each event into various tasks. Although one event must be completed before the next one is started, various tasks within an event can occur simultaneously. The time it takes to complete an event, therefore, depends on the number of days required to finish its longest task. Similarly, the total time it takes to finish a project is the sum of the times it takes to finish the individual events.

Assume that the critical path information for a major construction project has been stored in a data file in the computer. You have been asked to analyze this information so that your company can develop a bid on the project. Specifically, the management would like a summary report that lists each event along with the number of days for the shortest task in the event and the number of days for the longest task in that event. In addition, they would like the total project length computed in days and converted to weeks (five days per week).

The data file named PROJECT contains three integers per line. The first number is the event number, the second number is the task number, and the third number is the number of days required to complete the task. The data have been stored such that all the task data for event 1 is followed by all the task data for event 2, and so on. The task data for a particular event is also in order. You do not know ahead of time how many entries are in the file, but there is an upper limit of 98 total events, so a trailer signal is used to indicate the last line in the data file. The last line will contain a value of 99 for the event number and zeros for the corresponding task and days.

1. Problem Statement

Determine and print a project completion timetable.

2. Input/Output Description

Input — a data file named PROJECT that contains the critical path information for the events in the project

Output — a report

3. Hand Example

Use the following set of project data for the hand example:

Event Number	Task Number	Days
1	1	5
1	2	4
2	1	3
2	2	7
2	3	4
3	1	6
99	0	0

The corresponding report based on this data is as follows:

PROJECT COMPLETION TIMETABLE

EVENT NUMBER	TASK MINIMUM	TASK MAXIMUM
1	4	5
2	3	7
3	6	6

TOTAL PROJECT LENGTH = 18 DAYS
                     = 3 WEEKS 3 DAYS

4. Algorithm Development

Decomposition

Read project event data and print event information
Print summary information

To develop an algorithm, we look at the way we compiled the data in the hand example. Because all the data for the first event is together, we scan down the list of data for event 1 and keep track of the minimum number of days and maximum number of days. When we reach the task for event 2, we print the information related to event 1 and then repeat the process for event 2. We continue until we reach event 99, which is a signal that we do not have any more data. We print the information for the last event and then print the summary information using a total in which we accumulated the total maximum number of days for all the events.

Initial Pseudocode

Path: total ← 0
    Read event, task, days
    number ← event
    min ← days
    max ← days
    While not done do
        If same event then
            Update min and max
        Else
            Print number, min, max
            total ← total + max
            number ← event
            min ← days
            max ← days
        Read event, task, days
    Print number, min, max for last event
    Print total

As we refine the pseudocode, we need to be more specific about how we determine if we are done. We also need to add more details to the update for the minimum and maximum days for an event. Finally, when we print the total, we also need to compute and print the number of weeks.

Final Pseudocode

Path: total ← 0
    Read event, task, days
    number ← event
    min ← days
    max ← days

```
            If event = 99 then
                    done ← true
            Else
                    done ← false
            While not done do
                    If event = number then
                            If days < min then
                                    min ← days
                            If days > max then
                                    max ← days
                    Else
                            Print number, min, max
                            total ← total + max
                            number ← event
                            min ← days
                            max ← days
                    Read event, task, days
                    If event = 99 then
                            done ← true
            Print number, min, max for last event
            weeks ← total/5
            plus ← total − weeks × 5
            Print total, weeks, plus
```

Note that we need to do the computation for WEEKS so that the result is an integer instead of a real value.

FORTRAN Program

```
*-----------------------------------------------------------------*
        PROGRAM PATH
*
* This program determines the critical path
* information from project data stored in a data file.
*
        INTEGER TOTAL, EVENT, TASK, DAYS, NUMBER, MIN, MAX,
                WEEKS, PLUS
        LOGICAL DONE
*
        OPEN (UNIT=8,FILE='PROJECT',STATUS='OLD')
*
        PRINT*, 'PROJECT COMPLETION TIMETABLE'
        PRINT*
        PRINT*, 'EVENT NUMBER    TASK MINIMUM    TASK MAXIMUM'
*
        TOTAL = 0
*
        READ (8,*) EVENT, TASK, DAYS
        NUMBER = EVENT
        MIN = DAYS
        MAX = DAYS
*
```

```
              IF (EVENT.EQ.99) THEN
                 DONE = .TRUE.
              ELSE
                 DONE = .FALSE.
              END IF
     *
         5 IF (.NOT.DONE) THEN
              IF (EVENT.EQ.NUMBER) THEN
     *
                 IF (DAYS.LT.MIN) THEN
                    MIN = DAYS
                 ELSE IF (DAYS.GT.MAX) THEN
                    MAX = DAYS
                 END IF
     *
              ELSE
     *
                 PRINT 10, NUMBER, MIN, MAX
        10       FORMAT (1X,I6,11X,I6,11X,I6)
                 TOTAL = TOTAL + MAX
                 NUMBER = EVENT
                 MIN = DAYS
                 MAX = DAYS
     *
              END IF
              READ(8,*) EVENT, TASK, DAYS
              IF(EVENT.EQ.99) DONE = .TRUE.
              GO TO 5
           END IF
     *
           PRINT 10, NUMBER, MIN, MAX
           TOTAL = TOTAL + MAX
     *
           WEEKS = TOTAL/5
           PLUS = TOTAL - WEEKS*5
           PRINT 15, TOTAL
        15 FORMAT (/,1X,'TOTAL PROJECT LENGTH = ',I2,' DAYS')
           PRINT 20, WEEKS, PLUS
        20 FORMAT (1X,20X,' = ',I2,' WEEKS',I2,' DAYS')
     *
           END
     *-------------------------------------------------------------*
```

The data files used in the Application sections, along with the corresponding programs, are available on a diskette that is available to your instructor. Check with your instructor to see if this information has been stored on your computer system so that you can access these programs and data files.

5. Testing

If we use the sample set of data from the hand example with this program, the following report is printed:

```
PROJECT COMPLETION TIMETABLE

EVENT NUMBER        TASK MINIMUM         TASK MAXIMUM
      1                  4                    5
      2                  3                    7
      3                  6                    6


TOTAL PROJECT LENGTH = 18 DAYS
                     =  3 WEEKS 3 DAYS
```

4-4 Application   SONAR SIGNALS

Acoustical Engineering

The study of sonar (*so*und *n*avigation *a*nd *r*anging) includes the generation, transmission, and reception of sound energy in water. Oceanological applications include ocean topography, geological mapping, and biological signal measurements; industrial sonar applications include fish finding, oil and mineral exploration, and acoustic beacon navigation; naval sonar applications include submarine navigation and submarine tracking. An active sonar system transmits a signal that often is a cosine with a known frequency. The reflections or echoes of the signal are then received and analyzed to provide information about the surroundings. A passive sonar system does not transmit signals but collects signals from sensors and analyzes them based on their frequency content.

To test algorithms that analyze sonar signals, engineers and scientists would first work with simple simulated sonar signals instead of actual sonar signals. A simple sonar signal can be represented by the following equation:[1]

$$x(t) = \sqrt{\frac{2E}{PD}} \cos(2\pi f_c t), \qquad 0 \le t \le PD$$

$$= 0, \qquad\qquad\qquad\quad \text{elsewhere}$$

where E = the transmitted energy
   PD = the pulse duration in seconds
   f_c = the frequency in Hz

Durations of a sonar signal can range from a fraction of a millisecond to several seconds, and frequencies can range from a few hundred Hz (hertz) to tens of kHz (kilohertz), depending on the sonar system and its desired operating range.

Write a program in which the user inputs the values for E, PD, and f_c. Then generate a data file called SONAR that contains samples of this sonar signal. The sampling of the signal should cover the pulse duration and be

---

[1] A. B. Baggeroer, "Sonar Signal Processing," Chapter 6 of *Applications of Digital Signal Processing* (Englewood Cliffs, N.J.: Prentice-Hall, 1978).

such that there are ten samples for every period of $x(t)$, where a period is equal to $1/f_c$ seconds. The data file should contain two values for every line: a time value (starting at zero) and the corresponding value of the sonar signal.

1. Problem Statement

Generate a sonar signal that contains ten samples from each period of a specified cosine, covering a given time duration.

2. Input/Output Description

Input—the values of E (transmitted energy in joules), PD (pulse duration in seconds), and f_c (frequency in Hz)

Output—a data file named SONAR that contains time and signal values for the sonar pulse duration

3. Hand Example

For a hand example, we use the following values:

$$E = 500 \text{ joules}$$

$$PD = 0.5 \text{ milliseconds (ms)}$$

$$f_c = 3.5 \text{ kHz}$$

The period of the cosine is $1/3500$, or approximately 0.3 ms. Thus, to have 10 samples per period, the sampling interval must be approximately 0.03 ms. The pulse duration is 0.5 ms, and thus we need 17 samples of the following signal:

$$x(t) = \sqrt{\frac{2E}{PD}} \cos(2\pi f_c t)$$

$$= \sqrt{\frac{1000}{0.0005}} \cos(2\pi(3500)t)$$

$$= 1414.2 \cos(2\pi\, 3500\, t)$$

The values for the sonar signal are

t (ms)	$x(t)$
0.0	1414.2
0.03	1117.4

t (ms)	x(t)
0.06	351.7
0.09	−561.6
0.12	−1239.3
0.15	−1396.8
0.18	−968.1
0.21	−133.1
0.24	757.77
0.27	1330.6
0.30	1345.0
0.33	794.9
0.36	−88.8
0.39	−935.2
0.42	−1389.2
0.45	−1260.1
0.48	−602.1

4. Algorithm Development

Decomposition

Read energy, pulse duration, frequency.
Compute amplitude, sampling time, number of points.
Generate file containing time values and cosine values.

The problem description did not specify using a trailer signal or an initial record with the number of lines in the data file; therefore, we will generate a data file with only valid data lines.

Pseudocode

Pulse: Read energy (E), pulse duration (PD), frequency

$$A \leftarrow \sqrt{\frac{2E}{PD}}$$

$$\text{period} \leftarrow \frac{1}{\text{frequency}}$$

$$\text{sample interval} \leftarrow \frac{\text{period}}{10}$$

$$\text{number of samples} \leftarrow \frac{PD}{\text{sample interval}}$$

For k = 0 to (number of samples − 1) do
      time ← k · (sample interval)
      signal ← A cos(2π frequency · time)
      Write time and signal to data file

FORTRAN Program

```
*------------------------------------------------------------*
      PROGRAM PULSE
*
*  This program generates a sonar signal.
*
      INTEGER NUMBER, K
      REAL E, PD, FREQ, A, PERIOD, T, PI, TIME, SIGNAL
*
      OPEN (UNIT=10,FILE='SONAR',STATUS='NEW')
*
      PRINT*, 'ENTER TRANSMITTED ENERGY IN JOULES'
      READ*, E
      PRINT*, 'ENTER PULSE DURATION IN SECONDS'
      READ*, PD
      PRINT*, 'ENTER COSINE FREQUENCY IN HERTZ'
      READ*, FREQ
*
      A = SQRT(2.0*E/PD)
      PERIOD = 1.0/FREQ
      T = PERIOD/10.0
      NUMBER = NINT(PD/T)
      PI = 3.141593
*
      DO 10 K=0,NUMBER-1
         TIME = REAL(K)*T
         SIGNAL = A*COS(2.0*PI*FREQ*TIME)
         WRITE (10,*) TIME, SIGNAL
   10 CONTINUE
*
      END
*------------------------------------------------------------*
```

The nearest integer function (NINT) was used in the computation of the number of samples to round the answer.

5. Testing

The following are the terminal screen interaction and a plot of the SONAR file generated by MATLAB for a sample interaction:

Computer Output

```
ENTER TRANSMITTED ENERGY IN JOULES
500.0
ENTER PULSE DURATION IN SECONDS
0.005
ENTER COSINE FREQUENCY IN HERTZ
3500.0
```

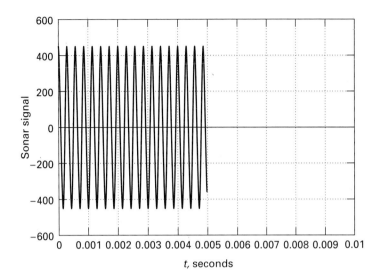

SUMMARY

Data files are necessary in engineering and scientific applications involving large amounts of data. This chapter presented the I/O statements for working with data files. It also discussed how to read information from data files. In particular, the chapter presented three techniques for reading data files that depend on whether there is a trailer signal or an initial line in the data file that contains the number of valid data records that follow it. We will use data files frequently throughout this module.

Key Words

data file	sentinel signal
END option	trailer signal
record	

Problems

This problem set begins with modifications to programs developed earlier in this chapter. Give the decomposition, pseudocode or flowchart, and FORTRAN program for each problem.

Problem 1 modifies the critical path analysis program PATH, given in Section 4.3.

1. Modify the critical path analysis program so that it prints the event and task number for all tasks requiring more than five days.

Problems 2–3 modify the sonar signals program PULSE, given in Section 4-4.

2. Modify the sonar signals program so that the output data file has a trailer signal with −99.0 as the first number on the trailer line.
3. Modify the sonar signals program so that it prints a first line in the data file that contains the number of valid data lines that follow.

In problems 4–9 develop programs using the five-step process.

4. As a practicing engineer, you have been collecting data on the performance of a new solar device. You have been measuring the sun's intensity and the voltage produced by a photovoltaic cell exposed to the sun. These measurements have been taken every 30 minutes during daylight hours for 2 months. Because the sun sets at a different time each day, the number of measurements taken each day may vary.

Each line of valid data contains the sun's intensity (integer), the time (24 hours, where 1430 represents 2:30 PM), and the voltage (real). A trailer signal contains 9999 for the sun's intensity. Write a complete program to read this data and compute and print the total number of measurements, the average intensity of the sun, and the average voltage value.

5. Write a program to compute fuel cost information for an automobile. The input data is stored in a data file called MILES. The first line in the file contains the initial mileage (odometer reading). Each following data line contains information collected as the automobile was refueled and includes the new mileage reading, the cost of the fuel, and the number of gallons of fuel. The last line of the file contains a negative value and two zeros, instead of the refueling information. The output of the program should be the cost per mile and the number of miles per gallon for each line of input data. Use the following report format:

```
                    FUEL COST  INFORMATION

   MILES        GALLONS     COST     COST/MILE    MILES/GALLON
   XXX.X        XX.X        XX.XX    X.XX         XX.X
```

6. Add the following summary information at the end of the report printed in problem 5:

```
                    SUMMARY  INFORMATION

          TOTAL MILES               XXXX.X
          TOTAL COST                XXXXX.XX
          TOTAL GALLONS             XXXX.X
          AVERAGE COST/MILE         X.XX
          AVERAGE MILES/GALLON      XX.X
```

7. Oil exploration and recovery is an important concern of large petroleum companies. Profitable oil recovery requires careful testing by drilling seismic holes and blasting with specific amounts of dynamite. For optimum seismic readings, a specific ratio between the amount of dynamite and the depth of the hole is required. Assume that each stick of dynamite is 2.5 feet long and weighs 5 pounds. The ideal powder charge requires a dynamite to depth-of-hole ratio of $1:3$. Thus, a 60-foot hole would require 20 feet of dynamite, which is equal to 8 sticks, or 40 pounds. The actual powder charge is not always equal to ideal powder charge because the ideal powder charge may not be in 5-pound increments; in these cases, the actual powder charge should be rounded down to the nearest 5-pound increment. (You cannot cut or break the

dynamite into shorter lengths for field operations.) The following example should clarify this process:

$$
\begin{aligned}
\text{Hole depth} &= 85 \text{ feet} \\
\text{Ideal charge} = 85/3 &= 28.33333 \text{ feet} \\
&= 11.33333 \text{ sticks} \\
&= 56.666666 \text{ pounds} \\
\text{Actual charge} &= 55 \text{ pounds} \\
&= 11 \text{ sticks}
\end{aligned}
$$

Information on the depths of the holes to be tested each day is stored in a data file called DRILL. The first line contains the number of sites to be tested that day. Each following line contains integer information for a specified site that gives the site identification number and the depth of the hole in feet. Write a complete program to read this information and print the report in the following format:

```
                     DAILY DRILLING REPORT

   SITE    DEPTH    IDEAL POWDER     ACTUAL POWDER     STICKS
   ID      (FT)     CHARGE (LBS)     CHARGE (LBS)
   12980   85       56.6666               55            11
```

8. Modify the program in problem 7 so that a final summary report follows the drilling report and has the form

```
   TOTAL POWDER USED = XXXXX LBS    (XXXX STICKS)
   TOTAL DRILLING FOOTAGE = XXXXXX FT
```

9. Modify the program in problem 8 so that it takes into consideration a special situation: If the depth of the hole is less than 30 feet, the hole is too shallow for blasting. Instead of printing the charge values for such a hole, print the site identification number, the depth, and the message HOLE TOO SHALLOW FOR BLASTING. The summary report should not include data for these shallow holes. Add a line to the summary report that contains the number of holes too shallow for blasting.

10. Assume that a data file called DATAXY contains a set of data coordinates that are to be used by several different programs. The first line of the file contains the number of data coordinates in the file, and each of the remaining lines contains the x and y coordinates for one of the data points. Since some of the programs need polar coordinates, we will generate a second data file that has each point in polar form, which is a radius and an angle in radians. Then, no matter how many programs use the data, it only has to be converted to polar coordinates once, and each program can then reference the appropriate file. Write a program to generate a new file called POLAR that contains the coordinates in polar form instead of rectangular form. The following equations convert a coordinate in rectangular form to polar form:

$$
r = \sqrt{x^2 + y^2} \qquad \theta = \arctan\left(\frac{y}{x}\right)
$$

(Be sure that the first line of the new data file specifies the number of data coordinates.)

11. Rewrite the program from problem 10, assuming that the original file is POLAR and that it contains data coordinates in polar form. The new output file should be called DATAXY and should contain data coordinates in rectangular form. The equations for converting polar coordinates to rectangular coordinates are

$$x = r\cos(\theta)$$
$$y = r\sin(\theta)$$

12. Rewrite the program from problem 10 such that it creates a data file called POINTS. The first line of data should still contain the number of coordinates in the data file. Each following line of the data file should contain four values. The first two values represent the rectangular coordinates (x and y), and the next two values represent the corresponding polar coordinates (r and θ) for the data point.

5   Array Processing

Wind Tunnels   To measure aerodynamic data for a new aircraft or a modification of an existing aircraft (such as a jumbo jet), an accurate scale model is mounted on a force-measuring support in a wind tunnel test chamber. The wind tunnel is then operated at different wind speeds, or Mach numbers (which are the wind speeds divided by the speed of sound), and measurements of the forces on the model are made at many different angles of the model relative to the wind direction. An aircraft can be tested at many positions, including those that give the pilot the most maneuverability but that also can be the most dangerous for actual flight testing. Nondestructive laser testing is also performed in wind tunnels to help determine which parts of the aircraft vibrate.

INTRODUCTION

This chapter develops a method for storing groups of values without explicitly giving each value a different name — each group (called an array) has a common name, but individual values have a unique index or subscript. This technique allows us to analyze the data using loops, where the common name remains the same but the index or subscript becomes a variable that changes with each pass through the loop. Because the data values are stored in separate memory locations, we can also access the data as often as needed without rereading it.

5-1   ONE-DIMENSIONAL ARRAYS

An *array* is a group of storage locations that have the same name. Individual members of an array are called *elements* and are distinguished by using the common name followed by a *subscript* or an index in parentheses. Subscripts are represented by consecutive integers, usually beginning with the integer 1. A *one-dimensional array* can be visualized as either one column of data or one row of data. The following diagram shows the storage locations and associated names for a one-dimensional integer array J of five elements and a one-dimensional real array DIST with four elements:

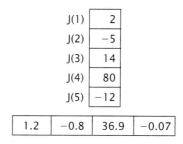

DIST(1) DIST(2) DIST(3) DIST(4)

Storage and Initialization

The DIMENSION statement, a nonexecutable statement, is used to reserve memory space or storage for an array. In its general form, a list of array names and their corresponding sizes follows the word DIMENSION, as shown:

> DIMENSION *array1(size), array2(size), . . .*

Array sizes must be specified with constants, not variables. A DIMENSION statement that reserves storage for the two arrays previously mentioned is

> DIMENSION J(5), DIST(4)

The number in parentheses after the array name gives the total number of values that can be stored in that array. Two separate DIMENSION statements, with one array listed in each statement, would also be valid but not preferable because it would require an extra statement. All DIMENSION statements must be placed before any executable statements in your program because they are specification statements.

The type of values stored in an array can be specified implicitly through the choice of array name or explicitly with a type specification statement. The following statement specifies that AREA is an array of 15 elements that contains integer values:

```
INTEGER AREA(15)
```

The typing of an array, whether implicit or explicit, applies to all elements of the array; hence, an array cannot contain both real values and integer values. Explicitly typed array names do not appear in DIMENSION statements because the array size has also been specified in the type statement. Because explicit typing is desirable, we will use explicit typing statements instead of DIMENSION statements for defining the arrays in the examples in this chapter.

The range of subscripts associated with an array can be specified with a beginning subscript number and an ending subscript number. Both numbers must be integers separated by a colon and must follow the array name in the DIMENSION statement or the type statement. The following statements reserve storage for a real array TAX whose elements are TAX(0), TAX(l), TAX(2), TAX(3), TAX(4), and TAX(5) and an integer array INCOME whose elements are INCOME(-3), INCOME(-2), INCOME(-1), INCOME(0), INCOME(l), INCOME(2), and INCOME(3):

```
REAL TAX(0:5)
INTEGER INCOME(-3:3)
```

Also note that the following declarations are equivalent:

```
INTEGER AREA(1:15)
INTEGER AREA(15)
```

Unless otherwise stated, we will assume in this module that all array subscripts begin with the integer 1. However, sometimes the range of the subscripts logically starts with an integer other than l. For example, if we have a set of population values for the years 1880–1980, it might be convenient to use the year to specify the corresponding population. We could specify such an array with the following statement:

```
INTEGER POPUL(1880:1980)
```

Then, if we wish to refer to the population for 1885, we use the reference POPUL(1885).

Values are assigned to array elements in the same way that values are assigned to regular variables. The following are valid assignment statements:

```
J(1) = 0
J(5) = NUM*COUNT
DIST(2) = 46.2 + SIN(X)
```

It is not valid to use an array name without a subscript in an assignment statement.

It is also extremely helpful to use variables and expressions, instead of constants, as subscripts. The following loop initializes all elements of the array J to the value 10. Observe that the variable I is used as a subscript and also as the DO loop index:

```
      DO 20 I=1,5
         J(I) = 10
   20 CONTINUE
```

The next loop initializes the array J to the values shown:

```
      DO 30 I=1,5
         J(I) = I
   30 CONTINUE
```

1	2	3	4	5

J(1) J(2) J(3) J(4) J(5)

The values of the array DIST are initialized to real values with this set of statements:

```
      DO 5 K=1,4
         DIST(K) = REAL(K)*1.5
    5 CONTINUE
```

1.5	3.0	4.5	6.0

DIST(1) DIST(2) DIST(3) DIST(4)

The previous examples illustrate that a subscript can be an integer constant or an integer variable. Subscripts can also be integer expressions, as indicated in the following statements:

```
      J(2*I) = 3
      R(J) = R(J-1)
      B = TR(2*I) + TR(2*I+1)
```

Whenever an expression is used as a subscript, be sure the value of the expression is between the starting and ending subscript value. If a subscript is outside the proper range, the program will not work correctly. With some compilers, a logic error message is given if a subscript is out of bounds; other compilers use an incorrect value for the invalid array reference, causing serious errors that are difficult to detect.

Input and Output

To read data into an array from a keyboard or from a data file, we use the READ statement. If we wish to read an entire array, we can use the name of the array without subscripts. We can also specify specific elements in a READ statement. If the array A contains three elements, then the following two READ statements are equivalent; if the array A contains eight elements, then the first READ statement reads values for all eight elements and the second READ statement reads values for only the first three elements:

```
      READ*, A
      READ*, A(1), A(2), A(3)
```

Array values may also be read using an *implied DO loop.* Implied DO loops use the indexing feature of the DO statement and may be used only in input and output statements and in the DATA statement presented in Section 5-2. For example, if we wish to read the first ten elements of the array R, we can use the following implied DO loop in the READ statement:

$$\text{READ*, } (R(I),I=1,10)$$

Further examples illustrate the use of these techniques for reading data into an array.

EXAMPLE 5-1

Temperature Measurements

A set of 50 temperature measurements has been entered into a data file, 1 value per line. The file is accessed with unit number 9. Give a set of statements to read this data into an array:

TEMP(l) ← line 1 of data file
TEMP(2) ← line 2 of data file
.
.
.
TEMP(50) ← line 50 of data file

SOLUTION 1

The READ statement in this solution reads one value, but it is in a loop that is executed 50 times and reads the entire array:

```
REAL TEMP(50)
   .
   .
   .
   DO 10 I=1,50
      READ (9,*) TEMP(I)
10 CONTINUE
```

SOLUTION 2

The READ statement in this solution contains no subscript; thus, it reads the entire array:

```
REAL TEMP(50)
   .
   .
   .
   READ (9,*) TEMP
```

SOLUTION 3

The READ statement in this solution contains an implied loop and is equivalent to a READ statement that lists TEMP(l), TEMP(2), . . ., TEMP(50):

```
REAL TEMP(50)
                .
                .
                .
READ (9,*) (TEMP(I),I=1,50)
```

Note that Solutions 2 and 3 are exactly the same as far as the computer is concerned; they both represent one READ statement with 50 variables. Solution 1 is the same as 50 READ statements with 1 variable per READ statement. If the data file contains 50 lines, each with 1 temperature measurement, all three solutions store the same data in the array TEMP.

However, suppose that each of the 50 lines in the data file has two numbers: a temperature measurement and a humidity measurement. Solution 1 will read a new data line for each temperature measurement because it is the equivalent of 50 READ statements. But, because Solutions 2 and 3 are the equivalent of one READ statement with 50 variables listed, they will go to a new line only when they run out of data. Thus, the data is stored as shown:

> TEMP(1) ← first temperature
> TEMP(2) ← first humidity
> TEMP(3) ← second temperature
> TEMP(4) ← second humidity
> .
> .
> .

This is a subtle, but important, distinction: The computer does not recognize that an error has occurred in the last example shown. It has data for the array and continues processing, assuming it has the correct data.

· ·

EXAMPLE 5-2   ## Mass Measurements

A group of 30 mass measurements is stored in a real array MASS. Print the values in the following tabulation:

```
MASS( 1) = XXX.X KG
MASS( 2) = XXX.X KG
        .
        .
        .
MASS(30) = XXX.X KG
```

SOLUTION

For each output line, we need to reference one value in the array. We can generate the values of the subscript with a DO statement that has an index of 1–30. The output form of this solution is important; go through it carefully to be sure you understand the placement of the literals in the FORMAT statement:

```
          REAL MASS(30)
            .
            .
            .
          DO 20 I=1,30
             PRINT 15, I, MASS(I)
       15    FORMAT (1X,'MASS(',I2,') = ',F5.1,' KG')
       20 CONTINUE
```

. . . .

Try It Try this self-test to check your memory of some key points from Section 5-1. If you have any problems with the exercises, you should reread this section. The solutions are given at the end of this module.

Problems 1–3 contain statements that initialize and print one-dimensional arrays. Show the output from each set of statements. Assume that each set of statements is independent of the others.

```
1.    INTEGER I, M(10)
      DO 5 I=1,10
         M(I) = 11 - I
    5 CONTINUE
      PRINT*, 'ARRAY VALUES:'
      DO 15 I=1,10
         PRINT 10, I, M(I)
   10    FORMAT (1X,'M(',I2,') = ',I4)
   15 CONTINUE

2.    INTEGER K, LIST(8)
      DO 5 K=1,8
         LIST(K) = K**2 - 4
    5 CONTINUE
      PRINT 10, LIST
   10 FORMAT (1X,8I4)

3.    INTEGER J
      REAL TIME(20)
      TIME(1) = 3.0
      DO 5 J=2,20
         TIME(J) = TIME(J-1) + 0.5
    5 CONTINUE
      PRINT 10, (J,TIME(J),J=1,20,4)
   10 FORMAT (1X,'TIME ',I2,' = ',F5.2)
```

5-2 DATA STATEMENT

The DATA statement is a specification statement and is therefore nonexecutable; it is useful in initializing both simple variables and arrays. The general form of the DATA statement is

> DATA *list of variable names/list of constants/*

An example of a DATA statement to initialize simple variables is

```
DATA SUM, VEL, VOLT, LENGTH /0.0,32.75,-2.5,10/
```

The number of data values must match the number of variable names. The data values should also be of the correct type so that the computer does not have to convert them. The preceding DATA statement initializes the following variables:

SUM	0.0

VEL	32.75

VOLT	−2.5

LENGTH	10

Because the DATA statement is a specification statement, it should precede any executable statements; it is therefore located near the beginning of your program, along with the REAL, INTEGER, LOGICAL, and DIMENSION statements. It follows these other statements because the specification of the types of variables or the declaration of an array should precede values given to the corresponding memory locations.

Caution should be exercised when using the DATA statement because it initializes values only at the beginning of program execution; this means that the DATA statement cannot be used in a loop to reinitialize variables. If it is necessary to reinitialize variables, you must use assignment statements.

If a number of values are to be repeated in the list of values, a constant followed by an asterisk indicates a repetition. The following statement initializes all four variables to zero:

```
DATA A, B, C, D /4*0.0/
```

A DATA statement can also be used to initialize one or more elements of an array:

```
INTEGER J(5)
REAL TIME(4)
DATA J, TIME /5*0,1.0,2.0,3.0,4.0/
```

J	0	0	0	0	0

TIME	1.0	2.0	3.0	4.0

```
REAL HOURS(5)
DATA HOURS(1) /60.0/
```

HOURS	60.0	?	?	?	?

The question marks indicate that some array elements were not initialized by the DATA statement. A syntax error would have occurred if the subscript was left off the array reference HOURS(l), because the number of variables would then not match the number of data values; that is, HOURS represents five variables, but HOURS(l) represents only one variable.

An implied loop can also be used in a DATA statement to initialize all or part of an array, as in

```
INTEGER YEAR(100)
DATA (YEAR(I),I=1,50) /50*0/
```

The first 50 elements of the array are initialized to zero, and the last 50 are not initialized. You must therefore make no assumptions about the contents of the last 50 values in the array.

5-3 Application   WIND TUNNELS

Aerospace Engineering
Wind tunnels are used to collect very accurate measurements of the lift and drag forces generated by a moving air stream on an aerodynamic body. When the data are subsequently used in engineering analyses, linear interpolation is frequently used to estimate values between the measured data points. This section explains linear interpolation and then uses linear interpolation in the analysis of the data from the wind tunnel.

Linear Interpolation
Interpolation is a numerical technique used to estimate unknown values of a function by using known values of the function at specific points. The simplest and probably the most popular interpolation method is linear interpolation. Essentially, all that is involved is connecting the data points with straight line segments and then interpolating for unknown values of the function on the straight line segments which connect adjacent data points. For example, consider the following graph that contains two values of a function at $x = 1$ and at $x = 3$. These two points have been connected with a straight line.

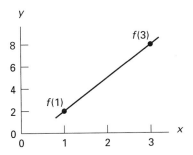

We can use linear interpolation to estimate the value of the function at

$x = 1.5$ using similar triangles, as shown in the following graph:

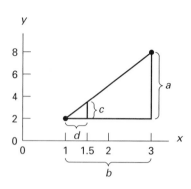

Thus,

$$\frac{a}{b} = \frac{c}{d}$$

or

$$\frac{f(3) - f(1)}{3 - 1} = \frac{f(1.5) - f(1)}{1.5 - 1}$$

If we solve this equation for $f(1.5)$, we have

$$f(1.5) - f(1) = \frac{(1.5 - 1) \cdot (f(3) - f(1))}{3 - 1}$$

or

$$\begin{aligned}
f(1.5) &= f(1) + \frac{(1.5 - 1) \cdot (f(3) - f(1))}{3 - 1} \\
&= 2 + \frac{(0.5) \cdot (8 - 2)}{2} \\
&= 3.5
\end{aligned}$$

We can develop a general linear interpolation formula for finding a function value $f(x)$ when x is between a and b and when the values of $f(a)$ and $f(b)$ are known. This formula is essentially the formula used in the previous example. It calculates the portion of the difference between the values on the x axis and then adds the corresponding portion of the difference between the values on the y axis to the leftmost function value:

$$f(x) = f(a) + \frac{(x - a)}{(b - a)} (f(b) - f(a))$$

This general formula works for linear interpolation of straight lines with either positive or negative slopes.

Computation of Lift and Drag Forces

In this section, we use a set of lift data on an aircraft wing. The 17 points in the data file are tabulated and plotted opposite to show the coefficient of lift, C_L, at various flight path angles of the wing for a Mach number of 0.3. It is easy to judge from the appearance of the data that linear interpolation provides a reasonably accurate method of estimating the lift coefficient at points between the data points, even in the region where the function is curved.

Flight Path Angle	Coefficient of Lift
−4	−0.202
−2	−0.050
0	0.108
2	0.264
4	0.421
6	0.573
8	0.727
10	0.880
12	1.027
14	1.150
15	1.195
16	1.225
17	1.244
18	1.250
19	1.245
20	1.221
21	1.177

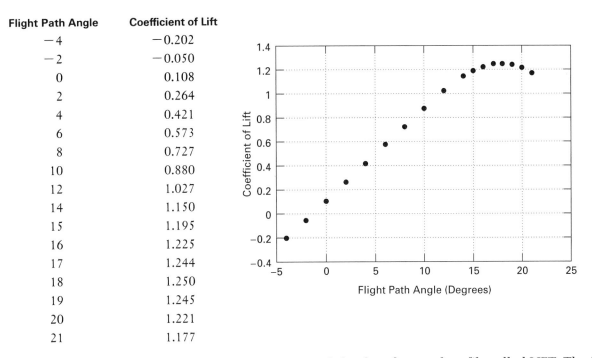

Write a program to read the data from a data file called LIFT. The first line in the data file contains the number of data pairs. Each successive line in the data file contains two numbers, where the first value represents alpha (the flight path angle in degrees), and the second value contains C_L (the coefficient of lift measured during the wind tunnel test for the corresponding angle). Assume that the data pairs contained in the data file are in ascending order based on alpha, with the smallest value of alpha first and the largest value of alpha last.

The program should then read a value of alpha entered from the keyboard for which we need to know the interpolated value of C_L. Using the algorithm shown earlier in this section, the program should interpolate for the lift coefficient, print the value of alpha and the corresponding estimate of C_L, and then prompt us for a new value of alpha. If we enter a value of alpha that is outside the range of the data, the program should print a message that the requested value is not within the interpolation region. The program terminates when we enter a value of 999.0 for alpha.

1. Problem Statement

Using linear interpolation with a wind tunnel data file, compute the lift coefficients that correspond to specified flight path angles.

2. Input/Output Description

Input — the wind tunnel data file and the values of alpha for which we need the interpolated values of C_L

Output — a list of the input values of alpha and the corresponding values of C_L

3. Hand Example

The data file that we use for testing contains the 17 data points given at the beginning of this section. Assume that we want to estimate the value of C_L that corresponds to alpha = 6.3. Using this data file and the linear interpolation formula from the previous section, the estimated value of C_L is

$$C_L = 0.573 + \frac{6.3 - 6}{8 - 6} (0.727 - 0.573)$$
$$= 0.573 + 0.15(0.154)$$
$$= 0.596$$

4. Algorithm Development

Decomposition

Read data file.
Read angles and interpolate coefficients.

As we refine these steps, remember that we need to determine if the input value of alpha is within the interpolation region. We can do this by checking to see if alpha is between the first flight path angle and the last flight path angle in the data file. If the input value of alpha is not within this allowable interpolation region, we print a message. We also need to check for an input alpha value of 999.0 so that we can terminate the program.

Pseudocode

Wind: Read N, the number of data pairs
     Read data file into arrays A and C
     Read input angle alpha
     While alpha ≠ 999.0 do
         If $A(1) \leq$ alpha $\leq A(N)$ then
             Find k such that $A(k - 1) \leq$ alpha $\leq A(k)$
             Interpolate for C_L

$$C_L \leftarrow C(k - 1) + \frac{\text{alpha} - A(k - 1)}{A(k) - A(k - 1)} \cdot (C(k) - C(k - 1))$$

             Print alpha, C_L
         Else
             Print message that alpha is out of range
         Read new input angle alpha

FORTRAN Program

```
*-------------------------------------------------------------*
      PROGRAM WIND
*
*  This program reads a file of wind tunnel test data and
*  then uses linear interpolation to compute lift
*  coefficients for additional flight angles.
*
      INTEGER N, K
      REAL A(500), C(500), ALPHA, CL
*
      OPEN (UNIT=10,FILE='LIFT',STATUS='OLD')
      READ (10,*) N
      DO 5 K=1,N
         READ (10,*) A(K), C(K)
    5 CONTINUE
*
      PRINT*, 'ENTER FLIGHT PATH ANGLE IN DEGREES'
      READ*, ALPHA
   10 IF (ALPHA.NE.999.0) THEN
*
         IF (A(1).LE.ALPHA.AND.ALPHA.LE.A(N)) THEN
*
            IF (ALPHA.EQ.A(1)) THEN
               K = 2
            ELSE
               K = 1
   15          IF (A(K).LT.ALPHA) THEN
                  K = K + 1
                  GO TO 15
               END IF
            END IF
            CL = C(K-1) + (ALPHA - A(K-1))/(A(K) - A(K-1))
     +                *(C(K) - C(K-1))
            PRINT 20, ALPHA, CL
   20       FORMAT (1X,'FLIGHT ANGLE = ',F7.2/
     +              1X,'CORRESPONDING COEFFICIENT OF ',
     +              'LIFT = ',F7.3)
*
         ELSE
*
            PRINT*,'ANGLE OUT OF RANGE FOR DATA FILE'
*
         END IF
*
         PRINT*
         PRINT*, 'ENTER FLIGHT PATH ANGLE IN DEGREES'
         PRINT*, '(999.0 TO QUIT)'
         READ*, ALPHA
         GO TO 10
      END IF
*
      END
*-------------------------------------------------------------*
```

5. Testing

We specified 500 values for arrays A and C, but these can be increased if we expect to need more values. Using the data file given at the beginning of this section, we tested the program with several values of alpha, including values outside the range of the data file. The interaction with the program is as follows:

```
ENTER FLIGHT PATH ANGLE IN DEGREES
6.3
FLIGHT ANGLE =    6.30
CORRESPONDING COEFFICIENT OF LIFT =   0.596

ENTER FLIGHT PATH ANGLE IN DEGREES
(999.0 TO QUIT)
20.5
FLIGHT ANGLE =   20.50
CORRESPONDING COEFFICIENT OF LIFT =   1.199

ENTER FLIGHT PATH ANGLE IN DEGREES
(999.0 TO QUIT)
-5
ANGLE OUT OF RANGE FOR DATA FILE

ENTER FLIGHT PATH ANGLE IN DEGREES
(999.0 TO QUIT)
22.5
ANGLE OUT OF RANGE FOR DATA FILE

ENTER FLIGHT PATH ANGLE IN DEGREES
(999.0 TO QUIT)
-4
FLIGHT ANGLE =   -4.00
CORRESPONDING COEFFICIENT OF LIFT =  -0.202

ENTER FLIGHT PATH ANGLE IN DEGREES
(999.0 TO QUIT)
21
FLIGHT ANGLE =   21.00
CORRESPONDING COEFFICIENT OF LIFT =   1.177

ENTER FLIGHT PATH ANGLE IN DEGREES
(999.0 TO QUIT)
10
FLIGHT ANGLE =   10.00
CORRESPONDING COEFFICIENT OF LIFT =   0.880

ENTER FLIGHT PATH ANGLE IN DEGREES
(999.0 TO QUIT)
999.0
```

Note that the first value computed by the program is the same value that we computed in the hand example. Also note that we tested all the special cases, including values of alpha that are out of range, values at the endpoints, and values at a data point.

5-4 SORTING ALGORITHMS

This section develops algorithms to *sort,* or reorder, a one-dimensional array or list in *ascending,* or low-to-high, order. (With minor alterations, the algorithm can be changed to one that sorts in *descending,* high-to-low, order. In Chapter 8 we use these sort algorithms to alphabetize character data.) The topic of sorting techniques is the subject of entire textbooks and courses; therefore, this module will not attempt to present all the important aspects of sorting. Instead, we present three common sorting techniques and develop pseudocode and FORTRAN solutions for all three so that you can compare the different techniques.

A *selection sort* is a simple sort that is based on finding the minimum value and placing it first in the list, finding the next smallest value and placing it second in the list, and so on. A *bubble sort* is based on interchanging adjacent values in the array until all the values are in the proper position. This sort is sometimes called a *multipass sort.* An *insertion sort* starts at the beginning of the list, comparing adjacent elements. If an element is out of order, it is continually exchanged with the value below it in the list until it is in its proper place. The sort then continues with the next element out of order.

No one sort algorithm is the best to use for all situations. To choose a good algorithm, we need to know something about the expected order of the data. For example, if the data are already very close to being in the correct order, both the insertion sort and the version of the bubble sort presented in this section are good choices. If the data are in a random order or close to the opposite order desired, then the insertion sort and the selection sort are good choices. None of these sorts is efficient if we are sorting a very large set of data. We should consult texts that cover other types of sorts in order to choose a good sort algorithm for a large set of data.

In the three sort algorithms presented in this section, we will use only one array. If we need to keep the original order of the data as well as the sorted data, we can copy the original data into a second array and sort it. Since the actual number of values in the array may be less than the maximum number of values that could be stored in it, we will assume that the variable COUNT specifies the actual number of data values to be sorted.

Selection Sort

We begin our look at the selection sort with a hand example. Consider the following list of data values:

Original List
4.1
7.3
1.7
5.2
1.3

In this algorithm, we first find the minimum value. Scanning down the list, we find that the last value, 1.3, is the minimum. We now want to put the value 1.3 in the first position of the array, but we do not want to lose the value 4.1 that is currently in the first position. Therefore, we will exchange the values. The switch of two values requires three steps, not two as you might imagine. Consider these statements:

$$X(I) = X(J)$$
$$X(J) = X(I)$$

Suppose X(I) contained the value 3.0 and X(J) contained the value − 1.0. The first statement will change the contents of X(I) from the value 3.0 to the value − 1.0. The second statement will move the value in X(I) to X(J), so that both locations contain − 1.0. These steps are shown next, along with the changes in the corresponding memory locations:

	X(I)	X(J)
	3.0	− 1.0
X(I) = X(J)	− 1.0	− 1.0
X(J) = X(I)	− 1.0	− 1.0

A correct way to switch the two values is shown here, along with the changes in the corresponding memory locations:

	X(I)	X(J)	HOLD
	3.0	− 1.0	?
HOLD = X(I)	3.0	− 1.0	3.0
X(I) = X(J)	− 1.0	− 1.0	3.0
X(J) = HOLD	− 1.0	3.0	3.0

Once we have switched the first value in the array with the value that has the minimum value, we search the values in the array from the second value to the last value for the minimum in that list. We then switch the second value with the minimum. We continue this until we are looking at the next-to-last and last values. If they are out of order, we switch them. At this point, the entire array will be sorted into an ascending order, as shown in the following diagram:

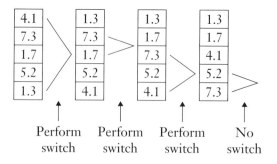

| Perform switch | Perform switch | Perform switch | No switch |

We now develop the pseudocode and FORTRAN statements for this selection sort.

Decomposition

> Sort list of data values into ascending order.

In the following refinement notice that we do not keep track of the minimum value itself; instead, we are interested in keeping track of the subscript or location of the minimum value. We need the subscript in order to be able to switch positions with another element in the array. Since the portion of the array that we search for the next minimum gets smaller, we use two variables, FIRST and LAST, to keep track of this array portion. FIRST will start at 2 and increment by 1 each time we do a switch. LAST will always be equal to COUNT since we search to the bottom of the list of valid values each time.

Pseudocode

(Assumes that values are already stored in the array X, and a variable COUNT specifies the number of valid data values in the array.)

```
Selection: last ← count
           For j = 1 to count − 1 do
               ptr ← j
               first ← j + 1
               For k = first to last do
                   If x(k) < x(ptr) then
                       ptr ← k
               Switch values in x(j) and x(ptr)
```

FORTRAN Statements

```
*
*   These statements sort the values in the array X
*   using a selection sort.  The variable COUNT
*   contains the number of valid data values in X.
*
      LAST = COUNT
      DO 10 J=1,COUNT-1
         PTR = J
         FIRST = J + 1
*
         DO 5 K=FIRST,LAST
            IF (X(K).LT.X(PTR)) PTR = K
    5    CONTINUE
*
         HOLD = X(J)
         X(J) = X(PTR)
         X(PTR) = HOLD
   10 CONTINUE
*
```

Bubble Sort

The basic step to the bubble sort algorithm is a single pass through the array, comparing adjacent elements. If a pair of adjacent elements is in the correct order (that is, the first value is less than or equal to the second value), we go to the next pair. If the pair is out of order, we switch the values and then go to the next pair.

The single pass through the array can be performed in a counting loop with index J. Each pair of adjacent values will be referred to by the subscripts J and J + 1. If the number of valid data elements in the array is stored in COUNT, we will make COUNT − 1 comparisons of adjacent values in a single pass through the array.

A single pass through a one-dimensional array, switching adjacent elements that are out of order, is not guaranteed to sort the values. Consider a single pass through the following array:

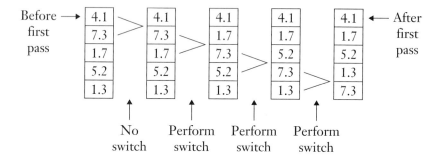

It will take two more complete passes before the array is sorted into ascending order, as shown in the following diagram:

After first pass		After second pass	
	4.1		1.7
	1.7		4.1
	5.2		1.3
	1.3		5.2
	7.3		7.3

After third pass		After fourth pass	
	1.7		1.3
	1.3		1.7
	4.1		4.1
	5.2		5.2
	7.3		7.3

A maximum of COUNT − 1 passes may be necessary to sort an array with this technique. If no switches are made during a single pass through the array, however, it is in ascending order. Thus, our algorithm for sorting a one-dimensional array will be to perform single passes through the array making switches until no elements are out of order. In developing the pseudocode, we use a logical variable SORTED that is initialized to true at the beginning of each pass through the data array. If any adjacent values are out of order, we switch the values and then change the value of SORTED to false because at least one pair of values was out of order on the pass. At the end of a pass through the data, if the value of SORTED is still true, the array is in ascending order.

We will make one more addition to the algorithm. During the first pass through the array, we switch any adjacent pairs that are out of order. Although this does not necessarily sort the entire array, it is guaranteed to move the largest value to the bottom of the list. During the second pass, the next-largest value will be moved to the next-to-the-last position. Therefore, with each pass we can reduce the number of positions that we check by 1.

Decomposition

> Sort list of data values into ascending order.

Pseudocode

(Assumes that the values are already stored in the array X, and a variable COUNT specifies the number of valid data values in the array.)

Bubble: sorted ← false
       first ← 1
       last ← count − 1
       While not sorted do
              sorted ← true
              For j = first to last do
                     If x(j) > x(j + 1) then
                            Switch values
                            sorted ← false
              last ← last − 1

FORTRAN Statements

```
*
*   These statements sort the values in the array X
*   using a bubble sort.  The variable COUNT
*   contains the number of valid data values in X.
*
      SORTED = .FALSE.
      FIRST = 1
      LAST = COUNT - 1
    5 IF (.NOT.SORTED) THEN
         SORTED = .TRUE.
*
         DO 10 J=FIRST,LAST
            IF (X(J).GT.X(J+1)) THEN
               HOLD = X(J)
               X(J) = X(J+1)
               X(J+1) = HOLD
               SORTED = .FALSE.
            END IF
   10    CONTINUE
*
         LAST = LAST - 1
         GO TO 5
      END IF
*
```

Insertion Sort

The insertion sort starts at the beginning of the list, comparing adjacent elements. If an element is out of order, we switch it with the previous element and then check to see if it is in its proper place. If not, we switch it with the new previous element and again check. We continue moving the element up in the array until it is in its proper position. We then return to the position in the list where we located the element out of order and pick up at that point, comparing the next pair of adjacent elements. When we reach the end of the list, it will be in order because each element that we found out of order was inserted in its proper position before we continued. The following diagram shows these steps with our sample array.

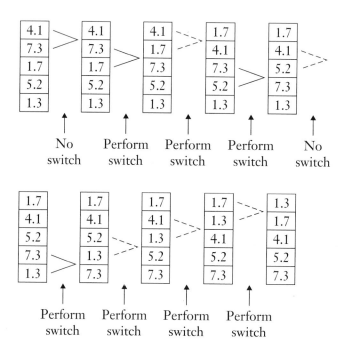

We now develop the pseudocode and FORTRAN statements for this insertion sort.

Decomposition

> Sort list of data values into ascending order.

In the following refinement, notice that we use the subscript j of the counting loop to point to our position before we begin backing up in the array to find the proper position for the element out of order. After putting the element in the correct spot, we can jump back to the next pair of elements in the list that we need to compare, since the value of the subscript j has not been changed.

Pseudocode

(Assumes that the values are already stored in the array X, and a variable COUNT specifies the number of valid data values in the array.)

Insertion: For j = 1 to count − 1 do
        If x(j) > x(j+1) then
            done ← false
            k ← j
            While not done do
                Switch x(k) with x(k+1)
                If (k = 1) or (x(k) ≥ x(k − 1)) then
                    done ← true
                Else
                    k ← k − 1

FORTRAN Statements

```
*
*   These statements sort the values in the array X
*   using an insertion sort.  The variable COUNT
*   contains the number of valid data values in X.
*
      DO 10 J=1,COUNT-1
*
         IF (X(J).GT.X(J+1)) THEN
*
            DONE = .FALSE.
            K = J
*
    5        IF (.NOT.DONE) THEN
                HOLD = X(K)
                X(K) = X(K+1)
                X(K+1) = HOLD
                IF (K.EQ.1) THEN
                    DONE = .TRUE.
                ELSE IF (X(K).GE.X(K-1)) THEN
                    DONE = .TRUE.
                ELSE
                    K = K - 1
                END IF
                GO TO 5
            END IF
*
         END IF
*
   10 CONTINUE
*
```

The final condition was separated into two separate conditions in the FORTRAN statements because it would be invalid to examine X(K − 1) if K has the value 1.

Try It Try this self-test to check your memory of certain key points from Section 5-4. If you have any problems with the exercise, you should reread this section. The solution is given at the end of this module. Consider the following list with six elements in it:

31
24
63
16
21
8

1. Show the sequence of changes that occur in the list if it is sorted using the bubble sort algorithm.

5-5 TWO-DIMENSIONAL ARRAYS

If we visualize a one-dimensional array as a single column of data, we can visualize a *two-dimensional array* as a group of columns, as illustrated:

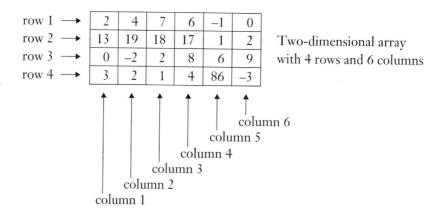

The diagram depicts an integer array with 24 elements. As in one-dimensional arrays, each of the 24 elements has the same array name. However, one subscript is not sufficient to specify an element in a two-dimensional array. For instance, if the array's name is M, it is not clear whether M(3) should be the third element in the first row or the third element in the first column. To avoid ambiguity, elements in a two-dimensional array are referenced with two subscripts: the first subscript references the row and the second subscript references the column. Thus, M(2,3) refers to the number in the second row, third column. In our diagram, M(2,3) would contain the value 18.

Storage and Initialization

Two-dimensional arrays must be specified with a DIMENSION statement or a type statement, but not both. The following type statements reserve storage for a one-dimensional real array B of ten elements, a two-dimensional array C with three rows and five columns, a two-dimensional real array NU with five rows and two columns, and a two-dimensional integer array J with seven rows and four columns:

```
REAL B(10), C(3,5), NUM(5,2)
INTEGER J(7,4)
```

This next statement reserves storage for a two-dimensional array with three rows and three columns:

```
REAL R(0:2,-1:1)
```

Two-dimensional arrays can be initialized with assignment statements, with input statements, and with the DATA statement. A two-dimensional array name can be used without subscripts in input statements, DATA statements, and output statements; if the name of the array is used in one of these statements without subscripts, the array is accessed in column order. In this module, we will explicitly use subscripts with two-dimensional arrays in order to be clear whether we are referencing the array elements in row order or in column order. It is common notation to use I for the row subscript and J for the column subscript.

EXAMPLE 5-3

Array Initialization, AREA

Define an array AREA with five rows and four columns. Fill it with the values shown:

1.0	1.0	2.0	2.0
1.0	1.0	2.0	2.0
1.0	1.0	2.0	2.0
1.0	1.0	2.0	2.0
1.0	1.0	2.0	2.0

SOLUTION 1

```
                REAL AREA(5,4)
                .
                .
                .
                DO 10 I=1,5
                    AREA(I,1) = 1.0
                    AREA(I,2) = 1.0
                    AREA(I,3) = 2.0
                    AREA(I,4) = 2.0
             10 CONTINUE
```

SOLUTION 2

```
            REAL AREA(5,4)
            DATA ((AREA(I,J),I=1,5),J=1,4) /10*1.0,10*2.0/
```

Array Initialization, SUM

Define and fill the integer array SUM as shown:

1	1	1
2	2	2
3	3	3
4	4	4

SOLUTION 1

If we observe that each element of the array contains its corresponding row number, then we can use the following solution:

```
INTEGER SUM(4,3)
  .
  .
  .
DO 10 I=1,4
   DO 5 J=1,3
      SUM(I,J) = I
 5    CONTINUE
10 CONTINUE
```

SOLUTION 2

The following DATA statement also initializes the array correctly:

```
INTEGER SUM(4,3)
DATA ((SUM(I,J),J=1,3),I=1,4) /3*1,3*2,3*3,3*4/
```

. . . .

EXAMPLE 5-5

Identity Matrix

A matrix is a structure used to store data that can be represented as a rectangular grid of numbers. Therefore, a matrix looks very similar to a two-dimensional array, except that a matrix has brackets on the sides while a two-dimensional array is shown inside a grid. When a matrix is used in a FORTRAN program, we store it in a two-dimensional array. When using matrix operations to solve engineering and science problems, we frequently use a matrix called an *identity matrix*. This matrix has the same number of rows as columns, so it is also called a *square matrix*. The identity matrix contains all 0's except for the main diagonal elements, which are 1's. (The main diagonal is composed of elements that have the same value for row number and for column number.) An identity matrix with five rows and five columns is shown.

$$\begin{bmatrix} 1 & 0 & 0 & 0 & 0 \\ 0 & 1 & 0 & 0 & 0 \\ 0 & 0 & 1 & 0 & 0 \\ 0 & 0 & 0 & 1 & 0 \\ 0 & 0 & 0 & 0 & 1 \end{bmatrix}$$

Define and fill a real array with these values.

SOLUTION

Because the value 1 appears at different positions in each row of the array, we cannot use the same type of solution we used in Example 5-4. If we list the elements that contain the value 1.0, we find that they are positions (1,1), (2,2), (3,3), (4,4), and (5,5); thus, the row and the column number are the same value. Recognizing this pattern, we can initialize the array as follows:

```
REAL IDEN(5,5)
    .
    .
    .
DO 10 I=1,5
    DO 5 J=1,5
        IF (I.EQ.J) THEN
            IDEN(I,J) = 1.0
        ELSE
            IDEN(I,J) = 0.0
        END IF
 5      CONTINUE
10 CONTINUE
```

. . . .

Input and Output

The main difference between using values from a one-dimensional array and using values from a two-dimensional array is that the latter requires two subscripts. Most loops used in the reading or printing of two-dimensional arrays are therefore nested loops. You can also use implied loops with two-dimensional array I/O. Avoid using only the array name on I/O statements with two-dimensional arrays; using the array name with no subscripts is equivalent to listing all the values in the array but in an order that goes down the columns instead of across the rows.

EXAMPLE 5-6 ## Reading Array Values from a Data File

Suppose that a data file contained temperature measurements taken at five times during an experiment. For each time, temperatures are taken at three locations in the jet engine being tested and stored on the same line in a data file. Read this data and store it in a real array TEMPS that has been dimensioned with five rows and three columns. Assume that the data file has already been opened.

CORRECT SOLUTION 1

This solution uses a single DO loop and lists the variable names for each row on the READ statement.

```
      DO 10 I=1,5
         READ (10,*) TEMPS(I,1), TEMPS(I,2), TEMPS(I,3)
   10 CONTINUE
```

CORRECT SOLUTION 2

This solution executes exactly like solution 1 because the implied DO loop on the READ is equivalent to listing the variables.

```
      DO 10 I=1,5
         READ (10,*) (TEMPS(I,J),J=1,3)
   10 CONTINUE
```

INCORRECT SOLUTION 1

This solution is the equivalent of 15 READ statements. Since each READ statement reads a new line in the data file, this solution tries to read each value from a new data line. Thus, the first number on line 1 will be read into TEMPS(1,1), the first number on line 2 will be read into TEMPS(1,2), and so on. The values are being stored in the wrong locations, and an execution error will occur when the program tries to read past the end of the data file.

```
      DO 10 I=1,5
         DO 5 J=1,3
            READ (10,*) TEMP(I,J)
    5    CONTINUE
   10 CONTINUE
```

INCORRECT SOLUTION 2

This solution is equivalent to one READ statement with all 15 variables written on it, but the variables are in column order, as in TEMPS(1,1), TEMPS(2,1), TEMPS(3,1), TEMPS(4,1), TEMPS(5,1), TEMPS(1,2), and so on. Since the order of the values in the data file is in row order, not column order, the values will be read into the wrong locations. Unfortunately, no error is detected by the computer, so we may not detect this error unless we carefully test our program.

```
      READ (10,*) TEMPS
```

EXAMPLE 5-7   ## Print Array Values

In this example we assume that we have correctly read the temperature data from the data file described in Example 5-6. We now want to print the data.

CORRECT SOLUTION 1

This solution uses a single DO loop and lists the variable names for each row on the PRINT statement. Five lines of output will be printed with three values per line, using a list-directed format.

```
      DO 10 I=1,5
         PRINT*, TEMPS(I,1), TEMPS(I,2), TEMPS(I,3)
   10 CONTINUE
```

CORRECT SOLUTION 2

This solution also prints three values per line using the specified format instead of the list-directed format.

```
      DO 10 I=1,5
         PRINT 3, TEMPS(I,1), TEMPS(I,2), TEMPS(I,3)
    3    FORMAT (1X,3(F5.2,2X))
   10 CONTINUE
```

INCORRECT SOLUTION

This solution prints one value per line instead of three because the format contains only one specification. When we run out of format specifications, we print the current buffer and back up in the format list to get a specification.

```
      DO 10 I=1,5
         PRINT 3, TEMPS(I,1), TEMPS(I,2), TEMPS(I,3)
    3    FORMAT (1X,F5.2)
   10 CONTINUE
```

. . . .

Try It

Try this self-test to check your memory of some key points from Section 5-5. If you have any problems with the exercises, you should reread this section. The solutions are given at the end of this module.

Problems 1–3 contain statements that initialize and print two-dimensional arrays. Draw the array and indicate the contents of each position in the array. Then, show the output from each set of statements. Assume that each set of statements is independent of the others.

```
1.    INTEGER I, J, CH(5,4)
      DO 10 I=1,5
         DO 5 J=1,4
            CH(I,J) = 2*(I + J)
    5    CONTINUE
   10 CONTINUE
      PRINT 15, (CH(5,J),J=1,4)
   15 FORMAT (1X,4I5)
```

```
2.    INTEGER I, J, K(3,3)
      DO 20 I=1,3
         K(I,1) = I
         K(I,2) = K(I,1) + 1
         K(I,3) = K(I,1) - 1
   20 CONTINUE
      DO 35 I=1,3
         PRINT 30, K(I,1), K(I,2), K(I,3)
   30    FORMAT (1X,3I4)
   35 CONTINUE
```

```
3.      INTEGER I, J
        REAL DIST(2,3), SUM
        SUM = 10.0
        DO 10 J=1,3
           DO 5 I=1,2
              SUM = SUM + 1.5
              DIST(I,J) = SUM
    5      CONTINUE
   10 CONTINUE
        DO 20 I=1,2
           PRINT 15, (DIST(I,J),J=1,3)
   15      FORMAT (1X,3F5.1)
   20 CONTINUE
```

5-6 Application   POWER PLANT DATA ANALYSIS

Power Engineering

The following table of data represents typical power outputs in megawatts from a power plant over a period of eight weeks. Each row represents one week's data; each column represents data taken on the same day
of the week. The data is stored one row per data line in a data file called
PLANT.

	Day 1	Day 2	Day 3	Day 4	Day 5	Day 6	Day 7
Week 1	207	301	222	302	22	167	125
Week 2	367	60	120	111	301	499	434
Week 3	211	62	441	192	21	293	316
Week 4	401	340	161	297	441	117	206
Week 5	448	111	370	220	264	444	207
Week 6	21	313	204	222	446	401	337
Week 7	213	208	444	321	320	335	313
Week 8	162	137	265	44	370	315	322

A program is needed to read the data, analyze it, and print the results in
the following composite report:

COMPOSITE INFORMATION
AVERAGE DAILY POWER OUTPUT = XXX.X MEGAWATTS
NUMBER OF DAYS WITH GREATER THAN AVERAGE POWER OUTPUT = XX
DAY(S) WITH MINIMUM POWER OUTPUT:
      WEEK X   DAY X
           .
           .
           .

1. Problem Statement

Analyze a set of data from a power plant to determine its average daily power output, the number of days with greater-than-average output, and the day or days that had minimum power output.

2. Input/Output Description

Input—8 weeks of daily power output stored in a data file

Output—a report summarizing the power output for the 8 weeks

3. Hand Example

For the hand example, we use a smaller set of data, but one that still maintains the two-dimensional array form. Consider this set of data:

	Day 1	Day 2
Week 1	311	405
Week 2	210	264
Week 3	361	210

First, we must sum all the values and divide by 6 to determine the average, which yields 1761/6, or 293.5 megawatts. Second, we compare each value to the average to determine how many values were greater than the average. In this small set of data, three values were greater than the average. Third, we must determine the number of days with minimum power output, which involves two steps: going through the data again to determine the minimum value and then going back through the data to find the day or days with the minimum power value. Then we can print its/their position(s) in the array. Using the small set of data, we find that the minimum value is 210 and that it occurred on two days. Thus, the output from our hand example is

COMPOSITE INFORMATION

AVERAGE DAILY POWER OUTPUT = 293.5 MEGAWATTS

NUMBER OF DAYS WITH GREATER THAN AVERAGE POWER OUTPUT = 3

DAY(S) WITH MINIMUM POWER OUTPUT:

WEEK 2   DAY 1

WEEK 3   DAY 2

4. Algorithm Development

Before we decompose the problem solution, it is important to spend some time considering the best way to store the data that we need for the program. Unfortunately, once we become comfortable with arrays, we tend to overuse them. Using an array complicates our programs because of the subscript handling. We should always ask ourselves, "should we really use an array for this data?"

If the individual data values will be needed more than once, an array is probably required. An array is also necessary if the data are not in the order needed. In general, arrays are helpful when we must read all the data before we can go back and begin processing it. However, if an average of a group of data values is all that is to be computed, we probably do not need an array; as we read the values, we can add them to a total and read the next value into the same memory location as the previous value. The individual values are not needed again because all the information required is now in the total.

Now, let us look at our specific problem and determine whether or not we need to use an array. First we need to compute an average daily power output. Then we need to count the number of days with output greater than average, which requires that we compare each output value to the average. For this application we need to store all the data in an array, and a two-dimensional array is the best choice of data structure.

We would like to minimize the number of passes through the data. We can compute the sum of the data points in the same loop in which we determine the minimum data value. However, we must make a separate pass through the data to determine how many values are greater than the average. Because the number of days with greater-than-average output is printed before we print the specific day or days that has/have minimum output, we need separate loops for these operations.

Decomposition

Read data.
Compute information.
Print information.

Initial Pseudocode

Powerplant: Read data
        Compute average power and minimum power
        Print average power
        Count days with above-average power
        Print count of days
        Print days with minimum power

Final Pseudocode

Powerplant: Read data
        Compute average and minimum value
        Print heading, average
        count ← 0
        For each data value do
                If data value > average then
                        count ← count + 1
        Print count
        For each data value do
                If data value = minimum then
                        Print row and column positions

FORTRAN Program

```
*------------------------------------------------------------------*
      PROGRAM PWRPLT
*
*  This program computes and prints a composite report
*  .summarizing eight weeks of power plant data.
*
      INTEGER TOTAL, COUNT, I, J, POWER(8,7), MIN
      REAL AVE
*
      DATA TOTAL, COUNT /0,0/
*
      OPEN (UNIT=12,FILE='PLANT',STATUS='OLD')
      DO 5 I=1,8
         READ (12,*) (POWER(I,J),J=1,7)
    5 CONTINUE
*
      MIN = POWER(1,1)
      DO 15 I=1,8
         DO 10 J=1,7
            TOTAL = TOTAL + POWER(I,J)
            IF (POWER(I,J).LT.MIN) MIN = POWER(I,J)
   10    CONTINUE
   15 CONTINUE
      AVE = REAL(TOTAL)/56.0
*
      DO 25 I=1,8
         DO 20 J=1,7
            IF (POWER(I,J).GT.AVE) COUNT = COUNT + 1
   20    CONTINUE
   25 CONTINUE
*
      PRINT 30, AVE, COUNT
   30 FORMAT (1X,15X,'COMPOSITE INFORMATION'/
     +        1X,'AVERAGE DAILY POWER OUTPUT = ',F5.1,
     +        ' MEGAWATTS'/
     +        1X,'NUMBER OF DAYS WITH GREATER THAN ',
     +        'AVERAGE POWER OUTPUT = ',I2)
*
      PRINT 45
   45 FORMAT (1X,'DAY(S) WITH MINIMUM POWER OUTPUT:')
      DO 60 I=1,8
         DO 55 J=1,7
            IF (POWER(I,J).EQ.MIN) PRINT 50, I, J
   50       FORMAT (1X,12X,'WEEK ',I2,'   DAY ',I2)
   55    CONTINUE
   60 CONTINUE
*
      END
*------------------------------------------------------------------*
```

5. Testing

This program should be tested in stages; again, the decomposition gives a good idea of the overall steps involved and can thus be used to identify the steps that should be tested individually. Remember that one of the most useful tools for debugging is the PRINT statement; use it to print the values of key variables in loops that may contain errors.

The output from this program using the data file given at the beginning of this section is

```
                  COMPOSITE INFORMATION
AVERAGE DAILY POWER OUTPUT = 259.2 MEGAWATTS
NUMBER OF DAYS WITH GREATER THAN AVERAGE POWER OUTPUT = 30
DAY(S) WITH MINIMUM POWER OUTPUT:
                WEEK  3    DAY  5
                WEEK  6    DAY  1
```

5-7 MULTIDIMENSIONAL ARRAYS

FORTRAN allows as many as seven dimensions for arrays. We can easily visualize a three-dimensional array such as a cube. We are also familiar with using three coordinates, X, Y, and Z, to locate points. This idea extends into subscripts. The following three-dimensional array could be defined with this statement:

REAL T(3,4,4)

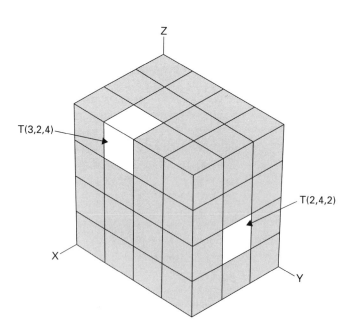

If we use the three-dimensional array name without subscripts, we access the array with the first subscript changing fastest, the second subscript changing second fastest, and the third subscript changing slowest. Thus, using the array T from the previous diagram, these two statements are equivalent:

```
READ *, T
READ *, (((T(I,J,K),I=1,3),J=1,4),K=1,4)
```

It should be evident that three levels of nesting in DO loops are often needed to access a three-dimensional array.

Most applications do not use arrays with more than three dimensions, probably because visualizing more than three dimensions seems too abstract. However, we now present a simple scheme that may help you to picture even a seven-dimensional array.

A four-dimensional array can be visualized as a row of three-dimensional arrays. The first subscript specifies a unique three-dimensional array. The other three subscripts specify a unique position in that array.

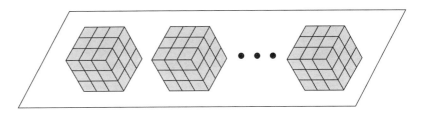

A five-dimensional array can be visualized as a block or grid of three-dimensional arrays. The first two subscripts specify a unique three-dimensional array. The other three subscripts specify a unique position in that array.

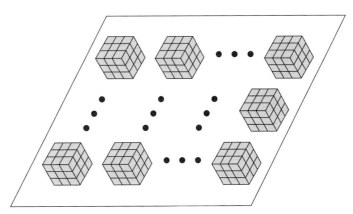

A six-dimensional array can be visualized as a row of blocks or grids. One subscript specifies the grid. The other five subscripts specify the unique position in the grid.

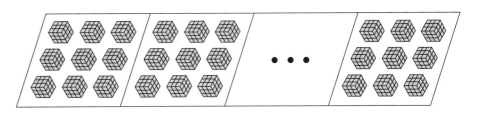

A seven-dimensional array can be visualized as a grid of grids or a grid of blocks. Two subscripts specify the grid. The other five subscripts specify the unique position in the grid.

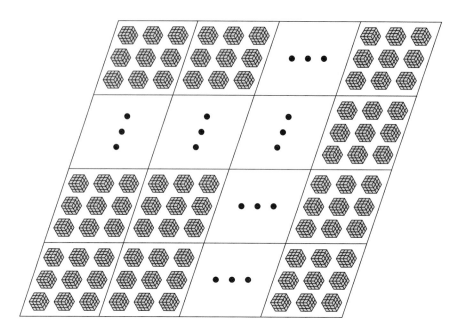

Now that you can visualize *multidimensional arrays,* a natural question is, "What dimension array do I use for solving a problem?" There is no single answer. A problem that can be solved with a two-dimensional array of four rows and three columns can also be solved with four one-dimensional arrays of three elements each. Usually the data fit one array form better than another; you should choose the form that is the easiest for you to work with in your program. For example, if you have census data from 10 countries over the period 1950–1980, you would probably use an array with 10 rows and 31 columns, or 31 rows and 10 columns. If the data represents the populations of 5 cities from each of 10 countries for the period 1950–1980, a three-dimensional array would be most appropriate; the three subscripts would represent year, country, and city.

SUMMARY

In this chapter you learned how to use an array—a group of storage locations that all have a common name but are distinguished by one or more subscripts. Arrays are one of the most powerful elements in FORTRAN because they allow you to keep large amounts of data easily accessible to your programs. The remaining chapters in this module rely heavily on arrays for storing and manipulating data. This chapter also presented three sort algorithms.

Key Words

array	bubble sort
ascending order	descending order

element
identity matrix
implied DO loop
insertion sort
multidimensional array
multipass sort

one-dimensional array
selection sort
sort
square matrix
subscript
two-dimensional array

Problems

This problem set begins with modifications to programs developed in this chapter. Give the decomposition, pseudocode or flowchart, and FORTRAN program for each problem. Problems 1 and 2 modify the wind tunnel analysis program WIND given in Section 5-3.

1. Modify the wind tunnel analysis program so that it prints the two flight path angles between which the maximum coefficient of lift occurs.

2. Modify the wind tunnel analysis program so that it asks the user to enter the flight path angle in degrees, but assume that the data file contains the angle in radians.

Problem 3 modifies the power plant data analysis program PWRPLT given in Section 5-6.

3. Modify the power plant data analysis program so that it reads a value N that determines the number of weeks that will be used for the report. Assume that N will never be more than 20.

For problems 4–6 assume that K, a one-dimensional array of 50 integer values, has already been filled with data.

4. Give FORTRAN statements to find and print the minimum value of K and its position or positions in the array in the following form:

```
MINIMUM VALUE OF K IS
K(XX) = XXXXX
```

5. Give FORTRAN statements to count the number of positive values, zero values, and negative values in K. The output form should be

```
XXX POSITIVE VALUES
XXX ZERO VALUES
XXX NEGATIVE VALUES
```

6. Give FORTRAN statements to replace each value of K with its absolute value, then print the array K with two values per line.

Develop these programs and program segments. Use the five-step process for all complete programs.

7. Give FORTRAN statements to print the last 10 elements of a real array M of size N. For instance, if M contains 25 elements, the output form is

```
M( 16) = XXX.X
M( 17) = XXX.X
        .
        .
        .
M( 25) = XXX.X
```

8. Give FORTRAN statements to interchange the first and one-hundredth elements, the second and ninety-ninth elements, and so on, of the array NUM that contains 100 integer values. See the diagram that follows:

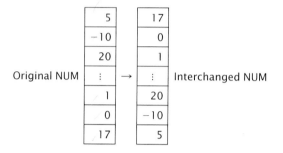

(*Hint:* You will need a temporary storage when you switch values.)

9. When a plot is made from experimental data, sometimes the scatter of the data points is such that it is difficult to select a "best representative line" for the plot. In such a case the data can be adjusted to reduce the scatter by using a "moving average" mathematical method of finding the average of three points in succession and replacing the middle value with this average.

Write a complete program to read an array Y of 20 real values from a file EXPR where the values are entered one per line. Build an array Z of 20 values where Z is the array of adjusted values. That is, Z(2) is the average of Y(1), Y(2), and Y(3); Z(3) is the average of Y(2), Y(3), and Y(4); and so on. Notice that the first and last values of Y cannot be adjusted and should be moved to Z without being changed. Do not destroy the original values in Y. Print the original and the adjusted values next to each other in a table.

10. A truck leasing company owns 12 delivery vans that are leased to several operators. Maintenance hours for each truck are allocated using a maintenance rate multiplied by the total number of hours accumulated by the entire fleet. The maintenance rate for each truck is determined from its percentage of the total fleet hours and the following table:

Percent of Hours	Maintenance Rate
0.00–9.99	0.02
10.00–24.99	0.04
25.00–100.00	0.06

A vehicle identification number and the monthly hours of use are entered in a data file HOURS. Write a complete program to read the data, convert each truck's hours to a percentage of the total fleet hours, and compute its maintenance hours using the maintenance rate. Round the calculated maintenance hours to the nearest hour, and ensure that each truck is allotted a minimum of one hour of maintenance each month. Print the following report:

```
MONTHLY MAINTENANCE REPORT
ID            HOURS         PERCENT       MAINTENANCE HOURS
XXX           XXXX          XXX.X         XXX
 .
 .
 .
TOTALS        XXXXX         XXX.X         XXXX
```

Test your program with the following input data:

ID	Hours
002	61
009	83
012	101
016	55
025	410
036	97
037	66
040	70
043	122
044	136
045	23
046	142

11. Assume that the reservations for an airplane flight have been stored in a file called FLIGHT. The plane contains 38 rows with 6 seats in each row. The seats in each row are numbered 1–6 as follows:

    1 Window seat, left side
    2 Center seat, left side
    3 Aisle seat, left side
    4 Aisle seat, right side
    5 Center seat, right side
    6 Window seat, right side

    The file FLIGHT contains 38 lines of information corresponding to the 38 rows. Each line contains 6 values corresponding to the 6 seats. The value for any seat is either 0 or 1, representing either an empty or an occupied seat.

    Write a complete program to read the FLIGHT information into a two-dimensional array called SEAT. Find and print all pairs of adjacent seats that are empty. Adjacent aisle seats should not be printed. If all three seats on one side of the plane are empty, then two pairs of adjacent seats should be printed. Print this information in the following manner:

```
              AVAILABLE SEAT PAIRS
                ROW         SEATS
                XX          X,X
                 .
                 .
                 .
                XX          X,X
```

If no pairs of seats are available, print an appropriate message.

12. Engineering data files often contain the dates on which information was recorded along with the information itself. If the data file is large, the date is often stored in a Julian date form, which is the year followed by the number of the day in the year (l to 365), since the Julian date will need only five digits, while a Gregorian date (month-day-year) requires six digits. For example, 010982 is a Gregorian date that converts to 82009 in Julian form. Write a complete program to convert a Gregorian date to a Julian date. Be sure to take leap years into account. (*Hint:* Use an array to store the number of days in each month.)

13. In problem 12 you saw that engineering data files sometimes contain Julian dates to minimize storage. However, when the information in the files is printed in reports, you want to convert the Julian dates to the more common Gregorian dates. Write a complete program to convert a Julian date to a Gregorian date.

6

Function Subprograms

Oil Well Production 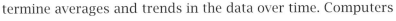Computer programs are used to analyze many types of information. In a large oil-producing company, they would be used to keep track of oil well production and to develop mathematical models for predicting oil well production. Computers would also be used to perform statistical analyses on the data to determine averages and trends in the data over time. Computers are also used to analyze results from experiments that collect data from seismometers (sensors that detect earth motion) and then attempt to model the geological structure under the ground. This information (often transmitted by satellite) is not only useful in predicting regions likely to contain oil, but it is also useful to engineers and scientists who are attempting to predict accurately the size and location of major earthquakes.

INTRODUCTION

As our programs become longer and more complicated, we find it harder to maintain program readability and simplicity. We also find that we frequently need to perform the same set of operations at more than one location in our programs. We can solve these problems by using *subprograms,* which are groups of statements that are defined separately and then referenced when we need them in our programs. FORTRAN has two types of subprograms: functions and subroutines. In this chapter, we review the intrinsic function (such as the square root function and the logarithm function) and learn how to write our own functions to perform computations unique to our applications. The function is quite useful in solving engineering and science problems because many of our solutions involve arithmetic computations. In the next chapter, we concentrate on subroutines which are subprograms that allow us to return many values as opposed to a single function value. Subroutines can also be used to read data files and print reports.

6-1 INTRINSIC FUNCTIONS

A *function* computes a single value, such as the square root of a number or the average of an array. You have already used functions in the form of intrinsic functions, such as SQRT and SIN. These intrinsic functions are in the compiler and are accessible directly from your program. The intrinsic functions (or *library functions*) available in FORTRAN 77 are listed in Appendix A. You should read through the list so that you are aware of the types of operations that can be performed with intrinsic functions. When you need to use one of these operations, you can refer to Appendix A for details on how to use that specific function. Chapter 2 introduced intrinsic functions, and the following list summarizes the main components of these functions:

1. The function name and its input values (or arguments) collectively represent a single value.
2. A function can never be used on the left side of an equal sign in an assignment statement.
3. The name of the intrinsic function typically determines the type of output from the function. (For example, if the name begins with one of the letters *I* through *N*, its value is an integer.)
4. The arguments of a function are generally of the same type as the function itself. For a few exceptions, refer to the list of intrinsic functions in Appendix A.
5. The arguments of a function must be enclosed in parentheses.
6. The arguments of a function may be constants, variables, expressions, or other functions.

Generic functions accept arguments of any allowable type and return a value of the same type as the argument. Thus, the generic function ABS will return an integer absolute value if its argument is an integer, but it will return a real absolute value if its argument is real. The table in Appendix A identifies generic functions.

6-2 STATEMENT FUNCTIONS

Engineering and science applications often require a function that is not included in the intrinsic function list. If the computation is needed frequently or requires several steps, we should implement it as a function instead of listing all the computations each time we need them. FORTRAN allows us to write our own functions in two ways: as a *statement function* or as a function subprogram. If the computation can be written in a single assignment statement, we can use the statement function that will be discussed in this section; otherwise, we must use the function subprogram that will be discussed in Section 6-4.

The general form for this statement is

> *function name (argument list) = expression*

The following rules apply to writing and using a statement function:

1. The statement function is defined at the beginning of the program, along with type statements and array definitions. It is a nonexecutable statement; thus, it should precede any executable statement.
2. The definition of the statement function contains the name of the function, followed by its arguments in parentheses, on the left side of an equal sign; the expression for computing the function value is on the right side of the equal sign.
3. The function name should be included in a type statement; otherwise, implicit typing will determine the function type.

Example 6-1 illustrates the use of the statement function in a complete program.

EXAMPLE 6-1   ## Triangle Area

The area of a triangle can be computed from the lengths of two sides and the angle between them:

$$AREA = 0.5*SIDE1*SIDE2*SIN(ANGLE)$$

Write a program that reads the lengths of the three sides of a triangle and the angles opposite each side.

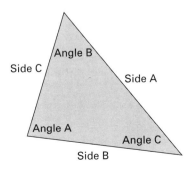

Compute and print the area of the triangle using one pair of sides and its corresponding angle. Then compute and print the area using another pair of sides and its corresponding angle. Finally, compute and print the area using the last pair of sides and its corresponding angle. Use a statement function to compute these areas.

SOLUTION

Step 1 is to state the problem clearly: Compute the area of a triangle.

Step 2 is to describe the input and output:

Input—lengths of the three sides a, b, c, and angles a, b, c
Output—areas computed from the sides and angles

Step 3 is to work a simple example. We choose a triangle with sides equal to 1.0, 1.0, $\sqrt{2}$ and angles 45°, 45°, and 90°. (The sides with length 1.0 are opposite the 45° angles.) Using this information, the three areas are

$$\text{area } 1 = 0.5 \cdot 1.0 \cdot 1.0 \cdot \sin(90°) = 0.5$$
$$\text{area } 2 = 0.5 \cdot 1.0 \cdot \sqrt{2} \cdot \sin(45°) = 0.5$$
$$\text{area } 3 = 0.5 \cdot 1.0 \cdot \sqrt{2} \cdot \sin(45°) = 0.5$$

Step 4 is to develop an algorithm, beginning with the decomposition.

Decomposition

Read the sides and angles of a triangle.
Compute the area in three ways.
Print the three areas.

Pseudocode

Triangle: Read sides a, b, c and angles a, b, c
Compute area using side b, side c, and angle a
Compute area using side a, side c, and angle b
Compute area using side a, side b, and angle c
Print three areas

FORTRAN Program

```
*---------------------------------------------------------------*
      PROGRAM TRIANG
*
*   This program reads the lengths of the sides of a triangle
*   along with the corresponding angles in radians. It then
*   computes and prints the area of the triangle using three
*   different sets of information.
*
      REAL SIDEA, SIDEB, SIDEC, A, B, C, AREA,
     +     AREAA, AREAB, AREAC, SIDE1, SIDE2, ANGLE
*
      AREA(SIDE1,SIDE2,ANGLE) = 0.5*SIDE1*SIDE2*SIN(ANGLE)
*
      PRINT*, 'ENTER THE LENGTHS OF THE THREE SIDES OF A'
      PRINT*, 'TRIANGLE IN THE FOLLOWING ORDER:'
```

```
      PRINT*, 'SIDE A    SIDE B    SIDE C'
      READ*, SIDEA, SIDEB, SIDEC
*
      PRINT*
      PRINT*, 'ENTER THE ANGLE OPPOSITE SIDE A,'
      PRINT*, 'THEN THE ANGLE OPPOSITE SIDE B,'
      PRINT*, 'AND THEN THE ANGLE OPPOSITE SIDE C.'
      PRINT*, '(IN RADIANS)'
      READ*, A, B, C
*
      AREAA = AREA(SIDEB,SIDEC,A)
      AREAB = AREA(SIDEC,SIDEA,B)
      AREAC = AREA(SIDEA,SIDEB,C)
*
      PRINT*
      PRINT*, 'THE THREE AREA COMPUTATIONS YIELD:'
      PRINT 5, AREAA, AREAB, AREAC
    5 FORMAT (1X,3F7.2)
*
      END
*-----------------------------------------------------------------*
```

Step 5 is to test the program. The following output represents a typical user interaction with this program:

```
ENTER THE LENGTHS OF THE THREE SIDES OF A
TRIANGLE IN THE FOLLOWING ORDER:
SIDE A    SIDE B    SIDE C
1.0    1.0    1.414

ENTER THE ANGLE OPPOSITE SIDE A,
THEN THE ANGLE OPPOSITE SIDE B,
AND THEN THE ANGLE OPPOSITE SIDE C.
(IN RADIANS)
0.785    0.785    1.571

THE THREE AREA COMPUTATIONS YIELD:
   0.50    0.50    0.50
```

Note that the arguments in the statement function definition are SIDEl, SIDE2, and ANGLE. These arguments do not represent variables used in our program; they tell the compiler that the statement function has three real arguments. When we referenced the function AREA to compute AREAA, note that the variable SIDEB corresponded to the argument SIDEl, the variable SIDEC corresponded to the argument SIDE2, and the variable C corresponded to the argument ANGLE. When we referenced the function AREA to compute AREAB and AREAC, different variables corresponded to the arguments SIDEl, SIDE2, and ANGLE.

· ·

Try It   Try this self-test to check your memory of some key points from Section 6-2. If you have any problems with the exercises, you should reread this section. The solutions are given at the end of this module.

For each of the following, give the statement function required to perform the computation.

1. Area of a square: $A = side^2$

2. Area of a parallelogram: $A = base \cdot height$

3. Area of a trapezoid: $A = \frac{1}{2} \cdot base \cdot (height1 + height\,2)$

6-3 MODULAR PROGRAMMING

In previous chapters, we stressed the importance of using While loops, DO loops, and IF structures as essential ingredients in writing structured programs. Another key element in structuring program logic is the use of *modules.* These modules or procedures allow us to write programs composed of nearly independent segments or routines. There are modules that are part of the language, such as intrinsic functions, and there are user-defined modules. In FORTRAN, user-defined modules are implemented as functions or subroutines. Functions and subroutines are also called subprograms; they look very much like programs except that they begin with a FUNCTION statement or a SUBROUTINE statement instead of a PROGRAM statement.

When we decompose a problem solution into a series of sequentially executed steps, we are decomposing the problem into steps that probably could be easily structured into functions and subroutines. The following are some important advantages to breaking programs into modules:

1. You can write and test each module separately from the rest of the program.
2. Debugging is easier because you are working with smaller sections of the program.
3. Modules can be used in other programs without rewriting or retesting.
4. Programs are more readable and thus more easily understood because of the modular structure.
5. Several programmers can work on different modules of a large program relatively independent of one another.
6. Individual parts of the program become shorter and therefore simpler.
7. A module can be used several times by the same program.

Since modules are so important in writing readable, well-structured programs, this chapter focuses on functions, and the next chapter focuses on subroutines.

The decomposition diagram shows the sequential steps necessary to solve a problem. Another type of diagram, the *structure chart,* is also very useful as we decompose our problem solution into smaller problems. While the decomposition diagram outlines the sequential operations needed, the structure chart outlines the modules but does not indicate the order in which they are to be executed.

The following diagram contains a structure chart for a program that we will develop in Section 6-5. In the program, we are reading oil well production data from a data file. For each oil well, we read the daily production in barrels, compute a daily average for the week, and print this information in the report. In addition, we keep summary information for all the wells and print it at the end of the report. A function subprogram is used to compute the average daily production for each oil well.

It is important to distinguish between a decomposition diagram and the structure chart. The decomposition diagram shows the sequential order of the steps of the solution; it does not identify any of the steps as modules. The structure chart, on the other hand, clearly defines the module definitions for the program but does not show the order in which the modules are used. Thus, both charts are useful in describing the algorithm that is being developed for the problem solution because they represent different information about the solution.

6-4   USER-DEFINED FUNCTIONS

Because intrinsic functions are contained in a library that is part of the compiler, you may find that a function in one computer manufacturer's compiler may not be available in another's. You may also find that you would like to use a function that is not a standard FORTRAN 77 intrinsic function. These problems can be solved by writing your own function.

A function subprogram, which is a program itself, is separate from the *main program*. It begins with a nonexecutable statement that identifies the function with a name and an argument list, as shown in the general form

FUNCTION *name (argument list)*

Because a function is separate from the main program, it must end with an END statement. The function should also contain a RETURN statement, which returns control to the statement that referenced the function. The general form of the RETURN statement is

RETURN

The rules for choosing a function name are the same as those for choosing a program name. In addition, the first letter of the function name specifies the type of value returned unless it is included in a specification statement.

The following statements illustrate a simple example with a structure chart, a main program, and a function subprogram.

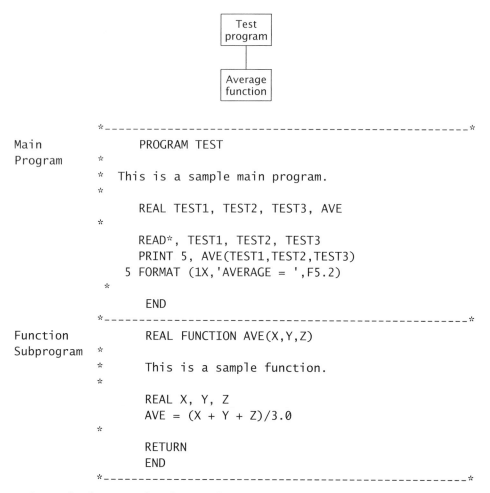

```
          *----------------------------------------------------------*
Main                PROGRAM TEST
Program       *
              *   This is a sample main program.
              *
                    REAL TEST1, TEST2, TEST3, AVE
              *
                    READ*, TEST1, TEST2, TEST3
                    PRINT 5, AVE(TEST1,TEST2,TEST3)
                  5 FORMAT (1X,'AVERAGE = ',F5.2)
               *
                    END
          *----------------------------------------------------------*
Function            REAL FUNCTION AVE(X,Y,Z)
Subprogram    *
              *     This is a sample function.
              *
                    REAL X, Y, Z
                    AVE = (X + Y + Z)/3.0
               *
                    RETURN
                    END
          *----------------------------------------------------------*
```

Several rules must be observed in writing a function subprogram.

1. The function arguments referenced in the main program are the *actual arguments.* They must match in type, number, and order the *dummy arguments* used in the FUNCTION statement. In the preceding example, the actual arguments are TEST1, TEST2, TEST3; the dummy arguments are X, Y, Z. TEST1 corresponds to X, TEST2 corresponds to Y, and TEST3 corresponds to Z.
2. If one of the arguments is an array, its dimensions must be specified in both the main program and the function subprogram.
3. The value to be returned to the main program is stored in the function name using an assignment statement.
4. When the function is ready to return control to the statement in the main program that referenced it, a RETURN statement is executed. A function may contain more than one RETURN statement.
5. A function can contain references to other functions, but it cannot contain a reference to itself.
6. A function subprogram is usually placed immediately after the main program, but it also may appear before the main program. If you have

more than one function, the order of the functions does not matter as long as each function is completely separate from the other functions.

7. A main program and its subprograms can be stored in the same program file or in separate files. If they are stored in separate files, it is necessary to link them before the main program can be executed. The statements required to perform the linking depend on the compiler and the operating system.

8. The same statement numbers may be used in both a function and the main program. No confusion occurs as to which statement is referenced because the function and the main program are completely separate. Similarly, a function and a main program can use the same variable name, such as SUM, to store different sums as long as the variable SUM is not an argument of the function.

9. The name of the function should appear in a type statement in the main program as well as in the function statement itself. The following statements illustrate the definition of an integer function AVE that computes an integer average. Compare this main program and function to the previous example that computes a real average.

```
*------------------------------------------------------------*
      PROGRAM TEST
*
*   This is a sample main program.
*
      INTEGER TEST1, TEST2, TEST3, AVE
*
      READ*, TEST1, TEST2, TEST3
      PRINT 5, AVE(TEST1,TEST2,TEST3)
    5 FORMAT (1X,'AVERAGE = ',I5)
*
      END
*------------------------------------------------------------*
      INTEGER FUNCTION AVE(I,J,K)
*
*   This is a sample function.
*
      INTEGER I, J, K
*
      AVE = (I + J + K)/3
*
      RETURN
      END
*------------------------------------------------------------*
```

10. Do not change the values of the dummy arguments. Some compilers return the new values to the main program; others do not. If you need to return more than the function value itself, use a subroutine (discussed in Chapter 7) instead of a function.

The following series of examples illustrates the use of these rules in writing and using user-defined functions. Because a function subprogram is a program separate from the main program, you should approach function design and implementation much as you would a complete program. Follow the same guidelines for developing a decomposition diagram, and then refine it using pseudocode or flowcharts until it is detailed enough to convert into FORTRAN. Because the function receives values through the argument list, the pseudocode or flowchart should begin with the name of the function and the variables that are the dummy arguments to the function.

As stated earlier, when arrays are used as arguments in a function, they must be dimensioned in the function subprogram as well as in the main program. Generally, the array should have the same size in the function as it does in the main program; however, when the size of an array is an argument to the subprogram, we use a technique called *variable dimensioning.* This technique allows us to specify an array of variable size in the subprogram. The argument value then sets the size of the array when the subprogram is executed. Example 6-2 illustrates the use of an array with a fixed size, and Example 6-3 illustrates the use of an array with a variable size. (Note that variable dimensioning refers to dimensioning of an array used as a dummy argument in a subprogram. Any array that is not a dummy argument, but that is defined and used in a subprogram, must be dimensioned in the subprogram with a constant because it is not dimensioned in the main program.)

EXAMPLE 6-2   ## Array Average, Fixed Array Size

Write a function that receives an array of 20 real values. Compute the average of the array and return it as the function value.

SOLUTION

Step 1 is to state the problem clearly: Write a function that computes the average of an array.

Step 2 is to describe the input and output:

> Input—array of 20 real values
> Output—array average

Step 3 is to work a simple example by hand. Let the input array contain the following values:

$$-2, 36, 24, -3, 19, 21, 8, 2, 5, 38, 16, -4, 2, 7, 17, 24, 9, -3, 6, 0$$

The average is then $\frac{222}{20} = 11.1$.

Step 4 is to develop an algorithm, starting with the decomposition.

Decomposition

Compute average of array x.
Return.

Pseudocode

Average(x): sum ← 0.0
              For i = 1 to 20
                    sum ← sum + x(i)
              average ← sum/20.0
              Return

FORTRAN Function

```
*------------------------------------------------------------------*
      REAL FUNCTION AVE(X)
*
*  This function computes the average of a real
*  array with twenty values.
*
      INTEGER I
      REAL X(20), SUM
*
      SUM = 0.0
      DO 10 I=1,20
         SUM = SUM + X(I)
   10 CONTINUE
      AVE = SUM/20.0
*
      RETURN
      END
*------------------------------------------------------------------*
```

Step 5 is to test the program. A portion of a main program that might use this function is

```
      INTEGER ID, I
      REAL SCORES(20), HWAVE, AVE
*
      READ*, ID
      READ*, (SCORES(I),I=1,20)
*
      HWAVE = AVE(SCORES)
```

. . . .

EXAMPLE 6-3

Median Value, Variable Array Size

The median of a list of sorted numbers is defined as the number in the middle. If the list has an even number of values, the median is defined as the average of the two middle values. Write a function called MEDIAN that has two dummy arguments: a real array and an integer that specifies the number of values in the array. The function should assume that the elements of the array have already been sorted. Return the median value of the array as the function value.

SOLUTION

Step 1 is to state the problem clearly: Write a function to determine the median value of an array.

Step 2 is to describe the input and output:

Input — a real array and an integer that specifies the number of values in the array

Output — median of the array

Step 3 is to work a simple example by hand. In the list $-5, 2, 7, 36$, and 42, the median is the number 7. In the list $-5, 2, 7, 36, 42$, and 82, the median is $(7 + 36)/2$, or 21.5.

Step 4 is to develop an algorithm, starting with the decomposition.

Decomposition

Determine median of array x.
Return.

We can use the MOD function to determine if the number of elements (n) in the array is odd or even. If n is odd, we must decide which subscript refers to the middle value. For example, if $n = 5$, we want the median to refer to the third value, which is referenced by $(n/2) + 1$. Recall that we are dividing two integers, and the result will be truncated to another integer. If n is even, we want to refer to the two middle values and compute their average. If $n = 6$, we want to use the third and fourth values, which can be referenced by $(n/2)$ and $(n/2) + 1$.

Pseudocode

Median(x,n): If n is odd, then

$$median \leftarrow x\left(\frac{n}{2} + 1\right)$$

Else

$$median \leftarrow \frac{x(n/2) + x(n/2 + 1)}{2}$$

Return

As we convert the pseudocode into FORTRAN, we must remember to specify that the function MEDIAN is a real function. (The default will be an integer.)

FORTRAN Function

```
*-------------------------------------/---------------------------------------*
      REAL FUNCTION MEDIAN(X,N)
*
*  This function determines the median value in a sorted
*  list of real numbers.
*
      INTEGER N
      REAL X(N)
*
      IF (MOD(N,2).NE.0) THEN
         MEDIAN = X(N/2+1)
      ELSE
         MEDIAN = (X(N/2) + X(N/2+1))/2.0
      END IF
*
```

```
      RETURN
      END
*----------------------------------------------------------------*
```

Step 5 is to test the program. This function could be tested by a program with the following structure chart and statements:

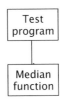

FORTRAN Program

```
*----------------------------------------------------------------*
      PROGRAM TEST
*
*  This program is written to test the median function.
*
      INTEGER N,I
      REAL X(10), MEDIAN
*
      PRINT*, 'ENTER NUMBER OF VALUES FOR ARRAY (<11)'
      READ*, N
      PRINT*, 'NOW ENTER SORTED ARRAY VALUES'
      READ*, (X(I),I=1,N)
      PRINT 5, MEDIAN(X,N)
    5 FORMAT (1X,'MEDIAN = ',F7.2)
*
      END
*----------------------------------------------------------------*
                MEDIAN function goes here
*----------------------------------------------------------------*
```

. . . .

EXAMPLE 6-4   ## Two-Dimensional Array Maximum

Write a function that will receive an array of integers with five rows and seven columns. The function should return the maximum value in the array.

SOLUTION

Step 1 is to state the problem clearly: Write a function that determines the maximum of a two-dimensional array.

Step 2 is to describe the input and output:

> Input — array with five rows and seven columns
> Output — maximum of input array

Step 3 is to work a simple example by hand. Let the array be the following:

$$\begin{array}{ccccccc} 2 & 5 & 3 & 1 & 6 & 2 & 19 \\ 20 & 32 & -4 & 2 & 7 & 13 & 2 \\ 8 & 25 & -5 & -4 & 18 & 4 & 9 \\ 28 & -4 & 2 & 6 & 7 & 2 & 5 \\ 3 & 8 & 13 & 23 & 5 & 3 & 9 \end{array}$$

The maximum value is 32.

Step 4 is to develop an algorithm, starting with the decomposition.

Decomposition

Determine maximum of array k.
Return.

To find the maximum value in the array, we initialize the maximum to the first value in the array; then we compare the rest of the elements in the array to the maximum, replacing the maximum value with any larger value that we find.

Pseudocode

Maxvalue(k): maxvalue $\leftarrow$ k(1,1)
        For i = 1,5 do
            For j = 1,7 do
                If k(i,j) > maxvalue then
                    maxvalue $\leftarrow$ k(i,j)
        Return

FORTRAN Function

```
*------------------------------------------------------------------*
      INTEGER FUNCTION MAXVAL(K)
*
*  This function determines the maximum value in an
*  integer array with 5 rows and 7 columns.
*
      INTEGER K(5,7), I, J
*
      MAXVAL = K(1,1)
      DO 10 I=1,5
         DO 5 J=1,7
            IF (K(I,J).GT.MAXVAL) MAXVAL = K(I,J)
    5    CONTINUE
   10 CONTINUE
*
      RETURN
      END
*------------------------------------------------------------------*
```

Step 5 is to test the program. A statement that might use the function in the main program, after filling an array NUMBER that has five rows and seven columns, is shown in the following group of statements:

```
      INTEGER NUMBER(5,7), MAXVAL
         .
         .
         .
      PRINT 5, MAXVAL(NUMBER)
    5 FORMAT (1X,'MAXIMUM NUMBER IS ',I5)
```

Problems can arise when you use variable dimensioning with arrays that have more than one dimension. The following is an example. First, we modify the MAXVAL function from Example 6-4 so that the number of rows and the number of columns are dummy arguments.

FORTRAN Function

```
*----------------------------------------------------------------*
      INTEGER FUNCTION MAXVAL(K,NR,NC)
*
*  This example is used to show problems that can occur
*  when using variable dimensioning with 2-D arrays.
*
      INTEGER NR, NC, K(NR,NC), I, J
*
      MAXVAL = K(1,1)
      DO 10 I=1,NR
         DO 5 J=1,NC
            IF (K(I,J).GT.MAXVAL) MAXVAL = K(I,J)
    5    CONTINUE
   10 CONTINUE
*
      RETURN
      END
*----------------------------------------------------------------*
```

If we modify the function reference to the one shown, the correct maximum value is printed:

```
      INTEGER NUMBER(5,7), MAXVAL
         .
         .
         .
      PRINT 5, MAXVAL(NUMBER,5,7)
    5 FORMAT (1X,'MAXIMUM NUMBER IS ',I5)
```

Suppose we want to determine the maximum value using only the first two rows and the first two columns of the array NUMBER, as shown in the following diagram:

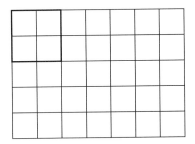

The following statements seem reasonable, but they are incorrect:

```
INTEGER NUMBER(5,7), MAXVAL
        .
        .
        .
     PRINT 5, MAXVAL(NUMBER,2,2)
   5 FORMAT (1X,'MAXIMUM NUMBER IS ',I5)
```

These statements are incorrect because FORTRAN stores a two-dimensional array by columns. The array NUMBER is actually stored in memory as shown in the following diagram:

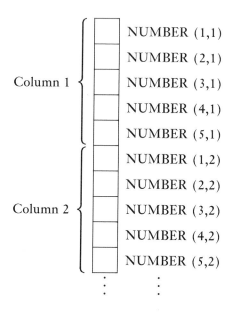

When we reference the array NUMBER in our function and specify that it has two rows and two columns, FORTRAN assumes that the values in NUMBER are stored in the following order:

Column 1 { NUMBER (1,1) / NUMBER (2,1) }
Column 2 { NUMBER (1,2) / NUMBER (2,2) }

When we put the preceding two diagrams side-by-side, we can see that the array elements do not completely match:

```
        ⎧ NUMBER (1,1) ⟷ NUMBER (1,1) ⎫ Column 1
        ⎪ NUMBER (2,1) ⟷ NUMBER (2,1) ⎭
Column 1 ⎨ NUMBER (3,1) ⟷ NUMBER (1,2) ⎫ Column 2
        ⎪ NUMBER (4,1) ⟷ NUMBER (2,2) ⎭
        ⎩ NUMBER (5,1)
```

In fact, if we reference an array with two rows and two columns using the array NUMBER, the elements we are actually going to use are outlined in the following diagram:

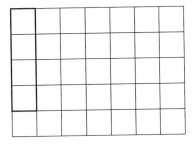

This problem does not result in a compiler error, but the values used by the function are not those that we intend it to use, which can present a difficult error to find in a program. In general, you should not use variable dimensions with arrays that have more than one dimension. If you have an application that requires variable dimensioning with two-dimensional arrays, include arguments that specify an operational size and a dimensional size, where the operational size specifies the part of the array you are actually going to use and the dimensional size specifies the size used in the array specification statement in the main program. We will use NR and NC for the size of the array we are going to use in the function and DR and DC for the dimensions used in the specification statement. These dummy arguments are used in the following function.

```
*-----------------------------------------------------------------*
      INTEGER FUNCTION MAXVAL(K,NR,NC,DR,DC)
*
*     This function determines the maximum value in an integer
*     array with NR rows and NC columns (original size DR by DC).
*
      INTEGER NR, NC, DR, DC, K(DR,DC), I, J
*
      MAXVAL = K(1,1)
      DO 10 I = 1,NR
         DO 5 J = 1,NC
            IF (K(I,J).GT.MAXVAL) MAXVAL = K(I,J)
    5    CONTINUE
   10 CONTINUE
*
      RETURN
      END
*-----------------------------------------------------------------*
```

Try It   Try this self-test to check your memory of some key points from Section 6-4. If you have any problems with the exercises, you should reread this section. The solutions are given at the end of this module.

For the following problems, give the value of S assuming that A = 8.9, B = 3.1, C = 0.2, and D = −5.5. The function TEST is the following:

```
*-----------------------------------------------------------------*
      REAL FUNCTION TEST(X,Y,Z)
*
*  This function returns either X+Y, X-Y, or X+Z
*  depending on the values of X, Y, Z.
*
      REAL X, Y, Z
*
      IF (X.GT.10.0) THEN
         TEST = X + Y
      ELSE IF (X.GT.Y) THEN
         TEST = X - Y
      ELSE
         TEST = X + Z
      END IF
*
      RETURN
      END
*-----------------------------------------------------------------*
```

1. S = TEST(A,B,C)

2. S = TEST(B,C,D)

3. S = TEST(ABS(D),C,D)

4. S = TEST(12.0,C,D*D)

6-5 Application   OIL WELL PRODUCTION

Petroleum Engineering

The daily production of oil from a group of wells is entered into a data file each week for analysis. One of the reports that uses this data file computes the average production from each well and prints a summary of the overall production from this group of wells.

Write a FORTRAN program that will read the information from the data file and generate this report. Assume that the first line of the data file contains a date (month, day, and year of the first day of the week that corresponds to the production data). Each following line in the data file contains an integer identification number for the well and seven real numbers that represent the well's production for the week. The number of wells to be analyzed varies from week to week, so a trailer line is included at the end of the file; this trailer contains the integer 99999, followed by seven zeros. You may assume that no well will have this integer as its identification number.

The report generated should have the following format:

```
         OIL WELL PRODUCTION
         WEEK OF XX-XX-XX

         WELL ID          AVERAGE PRODUCTION
                          (IN BARRELS)
         XXXXX            XXX.XX
```

Include a final line at the end of the report that gives the total number of oil wells plus their overall average.

Use the following set of data (stored in a file called WELLS) for testing the program:

```
05 06 92
52        87    136    0      54     60     82     51
63        54    73     88     105    20     21     105
24        67    98     177    35     65     98     0
8         23    34     52     67     180    80     3
64        33    55     79     108    118    130    20
66        40    44     63     89     36     54     36
67        20    35     76     87     154    98     80
55        10    13     34     23     43     12     0
3         34    56     187    34     202    23     34
2         98    98     87     34     54     100    20
25        29    43     54     65     12     15     17
18        45    65     202    205    100    99     98
14        36    34     98     34     43     23     9
13        0     9      8      4      3      2      10
36        23    88     99     65     77     45     35
38        23    100    134    122    111    211    0
81        23    34     54     98     5      93     82
89        29    58     39     20     50     30     47
99        100   12     43     98     34     23     9
45        23    93     75     93     2      34     8
88        23    301    23     83     23     9      20
77        28    12     43     43     92     83     98
39        98    43     12     23     54     23     98
12        43    54     92     84     75     72     91
48        83    138    189    73     27     49     10
99999     0     0      0      0      0      0      0
```

1. Problem Statement

Generate a report on oil well production from daily production data. Give the average production for individual wells and the overall average.

2. Input/Output Description

Input — a data file with daily production data for a group of oil wells
Output — a report summarizing the oil production

3. Hand Example

For the hand example, we use the first five lines of the data file given at the beginning of this application, along with the trailer line:

```
05 06 92
52      87    136    0     54    60    82    51
63      54    73     88    105   20    21    105
24      67    98     177   35    65    98    0
8       23    34     52    67    180   80    3
99999   0     0      0     0     0     0     0
```

We use the date in the first line of the data file in our heading. Each individual data line in the report contains the well identification (the first value in each line) followed by the average. We compute the average from the seven daily values that follow the identification. We continue computing the individual averages until we reach the trailer signal 99999. At this point, we average the individual wells and print a final summary line:

```
OIL WELL PRODUCTION
WEEK OF 5- 6-92
WELL ID                    AVERAGE PRODUCTION
                               (IN BARRELS)
         52                     67.14
         63                     66.57
         24                     77.14
          8                     62.71
OVERALL AVERAGE FOR 4 WELLS IS 68.39
```

4. Algorithm Development

Before we consider the steps in the algorithm, we should first decide the best way to store the data. For instance, do we want to store it in a two-dimensional array, or do we want to store the individual oil production amounts in a one-dimensional array, or can we avoid arrays altogether?

We answer these questions by looking at the way in which we need to use the data. To compute the individual well average, we need a sum of the individual production amounts. Because all seven values are entered in the same data line, we must read them at the same time; we need an array to store the seven daily production amounts for an individual well. To compute the overall average, we need a sum of the individual averages and a count of the number of wells; we do not need to keep the individual averages from each well.

In summary, we need a one-dimensional array to store the individual production values, and we do not need a two-dimensional array to store all the data.

We now decompose the algorithm into a sequence of steps. At the same time, we need to decide which operations or steps should be written as functions. Experience is the best guide to selecting which operations to define as functions. By studying how functions are used in the example programs, you will become more proficient at making these decisions.

Remember that there are often several ways in which functions can be used in a program.

Decomposition

Generate report.
Print summary line.

Initial Pseudocode

Report: Read date
        Read ID, oil production
        While ID ≠ 99999 do
            Determine individual average
            Print individual average
            Update overall average and well count
            Read next ID, oil production
        Print summary information

After completing the decomposition and the initial refinement, we can determine if any of the operations should be written as functions. Computations that are repeated in an algorithm and steps that involve long computations are good candidates for functions. Program readability suffers when the details of some of the steps become long and tedious. Even though the steps may be performed only once, placing them in a function may make the program simpler.

For the oil well production problem, a function can be used to compute the individual oil well averages. This operation is needed only once in the main loop, but it involves several steps. The main program will be more readable if the steps are moved to a function subprogram. The function to compute the individual averages is also one that would be useful in other programs that compute averages. To be flexible, we write the function assuming that the data is in an array whose size is one of the function arguments.

Final Pseudocode

Report: Read date
        number of wells ← 0
        total oil ← 0.0
        Read ID, oil production
        While ID ≠ 99999 do
            Compute individual average
            Print individual average
            Increment number of wells by 1
            Add individual well average to total oil
            Read next ID, production data
        Print summary information

Average(oil, n): sum ← 0.0
        For i = 1 to n
            sum ← sum + oil(i)
        average ← sum/n
        Return

Structure Chart

FORTRAN Program

```
*------------------------------------------------------------------*
      PROGRAM REPORT
*
*  This program generates a report from the daily
*  production information for a set of oil wells.
*
      INTEGER MO, DA, YR, ID, N, I
      REAL OIL(7), TOTAL, AVE, INDAVE
*
      DATA N, TOTAL /0,0.0/
*
      OPEN (UNIT=12,FILE='WELLS',STATUS='OLD')
*
      READ (12,*) MO, DA, YR
      PRINT*, 'OIL WELL PRODUCTION'
      PRINT 5, MO, DA, YR
    5 FORMAT (1X,'WEEK OF ',I2,'-',I2,'-',I2)
      PRINT*
      PRINT*, 'WELL ID       AVERAGE PRODUCTION'
      PRINT*, '                  (IN BARRELS)'
*
      READ (12,*) ID, (OIL(I),I=1,7)
   10 IF (ID.NE.99999) THEN
         INDAVE = AVE(OIL,7)
         PRINT 15, ID, INDAVE
   15    FORMAT (1X,I5,12X,F6.2)
         N = N + 1
         TOTAL = TOTAL + INDAVE
         READ (12,*) ID, (OIL(I),I=1,7)
         GO TO 10
      END IF
*
      PRINT*
      PRINT 20, N, TOTAL/REAL(N)
   20 FORMAT (1X,'OVERALL AVERAGE FOR ',I3,' WELLS IS ',F6.2)
*
      END
*------------------------------------------------------------------*
      REAL FUNCTION AVE(X,N)
*
*  This function computes the average of a real
*  array with N values.
*
      INTEGER N, I
      REAL X(N), SUM
```

```
*
      SUM = 0.0
      DO 10 I=1,N
          SUM = SUM + X(I)
   10 CONTINUE
*
      AVE = SUM/REAL(N)
*
      RETURN
      END
*------------------------------------------------------------------*
```

Could the variable SUM have been initialized in the function with a DATA statement? (The answer is no. Why?)

---

5. Testing

Begin testing this program using a small data set such as the one in the hand-worked example. The output from this program using the data file given at the beginning of this application is

```
OIL WELL PRODUCTION
WEEK OF  5- 6-92

WELL ID        AVERAGE PRODUCTION
                  (IN BARRELS)
    52              67.14
    63              66.57
    24              77.14
     8              62.71
    64              77.57
    66              51.71
    67              78.57
    55              19.29
     3              81.43
     2              70.14
    25              33.57
    18             116.29
    14              39.57
    13               5.14
    36              61.71
    38             100.14
    81              55.57
    89              39.00
    99              45.57
    45              46.86
    88              68.86
    77              57.00
    39              50.14
    12              73.00
    48              81.29

OVERALL AVERAGE FOR  25 WELLS IS  61.04
```

Could the last line of the data file contain only the trailer identification value 99999? (Are the seven zeros necessary in the last line of the data file?) If you try running the program without these last seven zeros, you will find that an execution error occurs because the program has run out of data. Because the identification number and the well production values are on the same line, we must read them with one READ statement. Each time the READ statement is executed, it reads eight values before control passes to the next statement. When the READ statement reaches the last line in the data file, it needs seven values in addition to the 99999 value before it can test for the trailer value.

6-6 FUNCTIONS FOR SEARCH ALGORITHMS

Solutions to different problems often include some of the same steps, such as computing the average of a set of values or finding the maximum of a set of values. Once we have written a function, such as one to compute the average of the values in an array, we can often reuse a function in another program with little or no modification to the function. Another very common operation performed with arrays is searching the array for a specific value. We may want to know if a particular value is in the array, how many times it occurs in the array, or where it first occurs in the array. All these forms of searches determine a single value and thus are good candidates for functions. *Searching algorithms* fall into two groups: those for searching an unordered list and those for searching an ordered list. In this section, we develop functions for both types of searches.

Unordered List

We first consider searching an unordered list; thus we assume the elements are not necessarily sorted into an ascending numerical order or any other order that may aid us in searching the array. The algorithm to search an unordered array is just a simple sequential search: check the first element, check the second element, and so on. There are several ways that we could implement this function. We could develop the function as an integer function that returns the position of the desired value in the array or zero if the desired value is not in the array. We could develop the function as an integer function that returns the number of times the element occurs in the array. We could also develop the function as a logical function that returns a value of true if the element is in the array or false if the element is not in the array. All these ideas represent valid functions, and we could think of programs that would use each of these forms. Since we have already written functions that return an integer value and functions that return a real value, we will implement this function as a logical function. We will call the function FOUND in order to make references to the function read smoothly. A reference to the function FOUND, as it might appear in a main program, is shown next.

```
IF (FOUND(X,COUNT,KEY)) THEN
    PRINT*, 'ITEM IS IN STOCK'
ELSE
    PRINT*, 'ITEM IS NOT IN STOCK'
END IF
```

Note that the function has three arguments, the array (X), the number of valid data values in the array (COUNT), and the value for which we are searching (KEY).

EXAMPLE 6-5 · Search Function for Unordered List

Write a logical function to search an unordered list for a specific value. The function should return a value of true if the specific value is found; otherwise, the function should return a value of false. The function will need three arguments: the array, the number of valid data entries in the array, and the value for which we are searching. Assume that all these values are integers.

SOLUTION

Step 1 is to state the problem clearly: Write a function to search an unordered list.

Step 2 is to describe the input and output:

Input—an array, the number of valid data entries in the array, and the value for which we are searching

Output—a logical value that indicates if we found the value

Step 3 is to work a simple example by hand. Suppose that the array contains the following values:

−7
21
14
8
52
77
15

If the value for which we are searching is 52, we look down the list and determine that the function should return a value of .TRUE. If the value for which we are searching is 53, then the function should return a value of .FALSE.

Step 4 is to develop an algorithm.

Decomposition

Search an unordered list for a specific value.

Pseudocode

```
Found(x,count,key): If count > 0 then
                              done ← false
                              i ← 1
                     Else
                              done ← true
                              found ← false
                     While not done do
                              If x(i) = key then
                                       done ← true
                                       found ← true
                              Else
                                       i ← i + 1
                              If i > count then
                                       done ← true
                                       found ← false
                     Return
```

FORTRAN Function

```
*--------------------------------------------------------------*
      LOGICAL FUNCTION FOUND(X,COUNT,KEY)
*
*  This function determines whether or not the key value
*  is in an unordered array.
*
      INTEGER COUNT, X(COUNT), KEY, I
      LOGICAL DONE
*
      IF (COUNT.GT.0) THEN
         DONE = .FALSE.
         I = 1
      ELSE
         DONE = .TRUE.
         FOUND = .FALSE.
      END IF
*
    5 IF (.NOT.DONE) THEN
*
         IF (X(I).EQ.KEY) THEN
            DONE = .TRUE.
            FOUND = .TRUE.
         ELSE
            I = I + 1
         END IF
*
         IF (I.GT.COUNT) THEN
            DONE = .TRUE.
            FOUND = .FALSE.
         END IF
*
         GO TO 5
      END IF
```

```
*
      RETURN
      END
*----------------------------------------------------------------*
```

Step 5 is to test the program. The following program could be used to test the search function. The program will ask the user to enter a set of data to be stored in the array. It will then ask the user to enter a specific value to be used in searching the array. The program will use the function to do the search; the program will then print a message giving the result of the search.

Structure Chart

FORTRAN Program

```
*----------------------------------------------------------------*
      PROGRAM TEST
*
*  This program tests the search function.
*
      INTEGER X(10), COUNT, KEY, J
      LOGICAL FOUND
*
      PRINT*, 'ENTER COUNT (<11) OF VALUES FOR LIST'
      READ*, COUNT
      PRINT*, 'ENTER VALUES'
      READ*, (X(J),J=1,COUNT)
      PRINT*, 'ENTER VALUE FOR SEARCH'
      READ*, KEY
*
      IF (FOUND(X,COUNT,KEY)) THEN
         PRINT*, KEY,' FOUND IN THE LIST'
      ELSE
         PRINT*, KEY,' NOT FOUND IN THE LIST'
      END IF
*
      END
*----------------------------------------------------------------*
                    FOUND function goes here
*----------------------------------------------------------------*
```

An example of a test of the function FOUND using this program is shown next:

```
ENTER COUNT (<11) OF VALUES FOR LIST
3
ENTER VALUES
26 -3 7
ENTER VALUE FOR SEARCH
15
    15 NOT FOUND IN THE LIST
```

. . . .

Ordered List

This section presents a common algorithm for searching an ordered list. The algorithm first checks the middle of the array and decides if the item for which we are searching is in the first half of the array or the second half of the array. If it is in the first half, we then check the middle of the first half and decide whether the item is in the first fourth of the array or the second fourth of the array. The process of dividing the array into smaller and smaller pieces continues until we find the element or find the position where it should have been. Since this technique continually divides the part of the array that we are searching in half, it is sometimes called a *binary search.*

EXAMPLE 6-6

Binary Search Function for Ordered List

In this example, we implement the search function using a binary search algorithm.

For a hand example, we first illustrate the binary search algorithm with the list in the margin in which we search for the value 25. There are seven values, the first referenced by a subscript value of 1 and the last referenced by a subscript value of 7. In a binary search, we compute the middle position by adding the first position number to the last position number and dividing by two. This should be done as an integer division. In our case, 7 plus 1 equals 8, and 8 divided by 2 is 4. Thus, we check the fourth position and compare its value to the value for which we are searching. The fourth value is 38, which is larger than 25, so we can narrow our search to the top half of the array. Our first position is still 1, and our last position is changed to the position above the middle position, or 3. We now divide that part of the array in half and compute the new midpoint to be $(1 + 3)/2$, or 2. The second value is 2, which is smaller than 25, so we can narrow our search to the second quarter of the array. Our first position is now one past the middle position, or 3, and our last position is 3. When the first and last positions are the same, we have narrowed in on the position where the value should be. Thus, we can exit the search algorithm. This specific example is illustrated in the following diagram. Follow through each step to be sure that you understand the sequence of steps needed.

−7
2
14
38
52
77
105

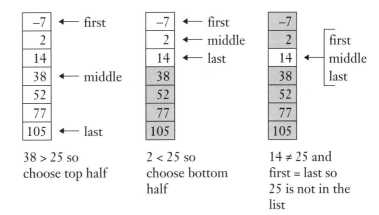

When the number of elements in the array is even, it is possible for the position of the first element to be greater than the position of the last element if the desired value is not in the list. Therefore, if the position of the first element is equal to or greater than the position of the last element, then the key value is not in the list.

The next step is to develop an algorithm. In the development of the function, we assume that the values are already stored in the array X, a variable COUNT specifies the number of valid data values in the array, and the variable KEY contains the value for which we are searching.

Decomposition

> Search an ordered list for a specific value.

Pseudocode

Found(x,count,key): If count $\leq$ 0 then
        found $\leftarrow$ false
    Else if x(1)>key or x(count)<key then
        found $\leftarrow$ false
    Else
        done $\leftarrow$ false
        found $\leftarrow$ false
        first $\leftarrow$ 1
        last $\leftarrow$ count
        While not done do
            middle $\leftarrow$ (first + last)/2
            If x(middle) = key then
                found $\leftarrow$ true
                done $\leftarrow$ true
            If not done then
                If first $\geq$ last then
                    done $\leftarrow$ true
                Else if x(middle) > key then
                    last $\leftarrow$ middle $-$ 1
                Else
                    first $\leftarrow$ middle + 1
    Return

FORTRAN Function

```
*------------------------------------------------------------------*
      LOGICAL FUNCTION FOUND(X,COUNT,KEY)
*
*  This function determines whether or not the key value
*  is in an ordered array using a binary search.
*
      INTEGER COUNT, X(COUNT), KEY, FIRST, LAST, MIDDLE
      LOGICAL DONE
*
      IF (COUNT.LE.0) THEN
*
         FOUND = .FALSE.
      ELSE IF (X(1).GT.KEY.OR.X(COUNT).LT.KEY) THEN
         FOUND = .FALSE.
*
      ELSE
*
         DONE = .FALSE.
         FOUND = .FALSE.
         FIRST = 1
         LAST = COUNT
*
    5    IF (.NOT.DONE) THEN
*
            MIDDLE = (FIRST + LAST)/2
*
            IF (X(MIDDLE).EQ.KEY) THEN
               FOUND = .TRUE.
               DONE = .TRUE.
            END IF
*
            IF (.NOT.DONE) THEN
               IF (FIRST.GE.LAST) THEN
                  DONE = .TRUE.
               ELSE IF (X(MIDDLE).GT.KEY) THEN
                  LAST = MIDDLE - 1
               ELSE
                  FIRST = MIDDLE + 1
               END IF
            END IF
*
            GO TO 5
         END IF
*
      END IF
*
      RETURN
      END
*------------------------------------------------------------------*
```

Although both the sequential search and the binary search correctly search for an item in an ordered list, the sequential sort is more efficient for small lists and the binary search is more efficient for large lists.

. . . .

SUMMARY

A function is a module that represents a single value. FORTRAN contains many intrinsic functions that compute values such as trigonometric functions and logarithms. We can also write our own functions. The statement function is a specification statement that defines a function within our program. A statement function can be used only when the function can be defined in a single assignment statement. A function subprogram is an independent module that is used to define functions that require more than a single assignment statement. All three forms of functions are useful in engineering and science applications because they simplify the computations in our programs. This chapter also presented several forms of functions for searching an array.

Key Words

actual argument	module
binary search	searching algorithm
dummy argument	statement function
function	structure chart
library function	subprogram
main program	variable dimensioning

Problems

This problem set begins with modifications to a program developed earlier in this chapter. Give the decomposition, pseudocode or flowchart, and FORTRAN solution for each problem.

Problems 1 and 2 modify the oil well production program REPORT given in Section 6-5.

1. Modify the oil production program so that it asks the user to enter the current date and then prints this date in the upper right-hand corner of the first page of the report. Thus, the report will show the date that the report was run in addition to the date of the time period for which the data was collected.

2. Using an additional function, modify the oil production program so that it determines the maximum daily production for each oil well. Print this value in addition to the well average.

For problems 3 and 4, write the statement function whose return value is described.

3. Volume of a rectangular parallelepiped:

$$V = \text{length} \cdot \text{width} \cdot \text{height}$$

4. Volume of a right circular cylinder:

$$V = \pi \cdot \text{radius}^2 \cdot \text{height}$$

For problems 5–9, write the function subprogram whose return value is described. The input to the function is an integer array K of 100 elements.

5. MAXI(K), the maximum value of the array K.
6. MINI(K), the minimum value of the array K.
7. NPOS(K), the number of values greater than or equal to zero in the array K.
8. NNEG(K), the number of values less than zero in the array K.
9. NZERO(K), the number of values equal to zero in the array K.

For problems 10–13, assume that you have a function subprogram DENOM, with input value x, to compute the following expression:

$$x^2 + \sqrt{1 + 2x + 3x^2}$$

Give the main program statements that use this function to compute and print each of the following expressions:

10. ALPHA $= \dfrac{6.9 + y}{y^2 + \sqrt{1 + 2y + 3y^2}}$

11. BETA $= \dfrac{\sin y}{y^4 + \sqrt{1 + 2y^2 + 3y^4}}$

12. GAMMA $= \dfrac{2.3z + z^4}{z^2 + \sqrt{1 + 2z + 3z^2}}$

13. DELTA $= \dfrac{1}{\sin^2 y + \sqrt{1 + 2\sin y + 3\sin^2 y}}$

In problems 14–19, develop these programs and functions.

14. Write a function whose input is a two-digit integer. The function is to return a two-digit number whose digits are reversed from the input number. Thus, if 17 is the input to the function, 71 is the output.

15. Write a function FACT that receives an integer value and returns the factorial of the value. Recall that the definition of a factorial is

$$0! = 1$$
$$n! = n(n - 1)(n - 2) \cdots 1$$

If $n < 0$, the function should return a value of zero.

16. The cosine of an angle may be computed from this series, where x is measured in radians:

$$\cos x = 1 - \frac{x^2}{2!} + \frac{x^4}{4!} - \frac{x^6}{6!} + \cdots$$

Write a function COSX whose input is an angle in radians. The function should compute the first 10 terms of the series and return that approximation of the cosine. Use the factorial function developed in problem 15. (Hint: The alternating sign can be obtained by computing $(-1)^K$. When K is even, $(-1)^K$ is equal to $+1$; when K is odd, $(-1)^K$ is equal to -1.)

17. Rewrite the function in problem 16 such that it computes the cosine with only as many terms of the series as are necessary to ensure that the absolute value of the last term is less than 0.000001.

18. Write a main program that will produce a table with three columns. The first column (x) should contain angles from 0.0 to 3.1 radians in increments of 0.1 radians. The second column should contain the cosines of the angles as computed by the intrinsic function. The third column should contain the cosines as computed by the function in problem 16. Print the cosine values with an F10.7 format.

19. Engineering programs often utilize experimental data that has been collected and then stored in tabular form. The programs read the data and use it to generate plots, displays, or additional data. Examples of such data include wind tunnel force data on a new aircraft design, rocket motor thrust data, or automobile engine horsepower and torque data. The table of data below gives the thrust output of a new rocket motor as a function of time from ignition.

Time	Thrust
0.0	0.0
1.0	630.0
2.0	915.0
5.0	870.0
8.0	860.0
12.0	885.0
13.0	890.0
15.0	895.0
20.0	888.0
22.0	860.0
23.0	872.0
26.0	810.0
30.0	730.0
32.0	574.0
33.0	217.0
34.0	0.0

Write a function called THRUST that computes the motor thrust for a rocket flight simulation program. The function should estimate the thrust for a specified time value using data from the preceding table. If the specified time value is not one of the times in the table, then the new data value can be determined from the tabulated values by using linear interpolation, which is also called linear proportional scaling. For example, the estimated thrust at $T = 2.2$ seconds would be computed in the following manner:

$$\text{THRUST} = 915.0 + \frac{2.2 - 2.0}{5.0 - 2.0} \times (870.0 - 915.0)$$

$$= 912.0$$

Assume that the function has two arguments: the two-dimensional real array with 16 rows and 2 columns, and the new value of time. If the time matches a time entry in the table, then the function should return the corresponding thrust value. If the time value does not match a time entry in the table, then the function should interpolate for a corresponding thrust value. Assume that the function will not be referenced if the time is less than zero or greater than 34.0.

7

Subroutine Subprograms

Flight Simulators Flight simulators can imitate all aspects of flight in a jumbo jet airliner. Multiple computers are used in a modern simulator to create the motion, vision, and sound cues associated with a flight between a number of specific cities. These systems are accurate enough for most training and routine checking of crew performance. The simulator is also considerably cheaper to use for training than a real airliner, and can be used to imitate a variety of in-flight emergencies that cannot be tested in actual flight. The software that controls the various aspects of the flight simulator uses randomly generated numbers so that each "trip" between two cities has differences in wind speeds, weather conditions, and in-flight emergencies.

INTRODUCTION

FORTRAN supports two types of subprograms: functions and subroutines. In Chapter 6 we developed programs using intrinsic functions, statement functions, and user-defined functions. Whereas a function is restricted to representing a single value, subroutines can compute more than one value and they are not limited to computing values. In this chapter, we will write subroutines and develop solutions to several applications using subroutines. We will use both functions and subroutines frequently in the remainder of this module to illustrate their importance in making programs simpler and more readable.

7-1 USER-DEFINED SUBROUTINES

Subroutines are modules written to perform operations that cannot be performed by a function. For example, if several values need to be returned from a module, a subroutine is used. A subroutine is also used for operations that do not compute values, such as reading the values in a data file. All subroutines in FORTRAN 77 are user-defined subroutines; there are no intrinsic subroutines.

Many of the rules for writing and using subroutines are similar to those for functions. The following list of rules outlines the differences between subroutines and functions.

1. A subroutine does not represent a value; thus, its name should be chosen for documentation purposes and not to specify a real or integer value.
2. A subroutine is referenced with an executable statement whose general form is

> CALL *subroutine name (argument list)*

3. The first line in a subroutine identifies it as a subroutine and includes the name of the subroutine and the argument list, as shown in this general form:

> SUBROUTINE *name (argument list)*

4. A subroutine uses the argument list not only for inputs to the subroutine but also for all values returned to the calling program. The subroutine arguments used in the CALL statement are the actual arguments, and the arguments used in the SUBROUTINE statement are the dummy arguments. The arguments in the CALL statement must match in type, number, and order those used in the subroutine definition.
5. A subroutine may return one value, many values, or no value. Similarly, a subroutine may have one input value, many input values, or no input value.
6. Because the subroutine is a separate program, the arguments are the only link between the main program and the subroutine. Thus, the choice of subroutine statement numbers and variable names is independent of those in the main program. The variables used in the sub-

routine that are not subroutine arguments are local variables, and their values are not accessible from the main program.

7. Be especially careful using multidimensional arrays in subroutines. It is generally advisable to pass both the dimensioned size and the operational size for arrays with two or more dimensions. You may want to review the discussion of this topic on page 169 in Chapter 6.

8. The subroutine, like the function, requires a RETURN statement to return control to the main program or to the subprogram that called it. It also requires an END statement because it is a complete program module.

9. In a flowchart, the following special symbol is used to show that the operations indicated are performed in a subroutine:

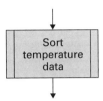

10. A subroutine may reference other functions or call other subroutines, but it cannot call itself.

The following two examples develop subroutines.

EXAMPLE 7-1   Array Statistics

Information commonly needed from a set of data includes the average, the minimum value, and the maximum value. These values can be computed using three functions; however, if all three are required, it is more efficient to compute them in one subprogram. Write a subroutine that is called with the statement

$$\text{CALL STAT}(X,N,XAVE,XMIN,XMAX)$$

where N is the number of valid elements in the real array X.

SOLUTION

Step 1 is to state the problem clearly: Write a subroutine to determine the average, minimum, and maximum of an array of values.

Step 2 is to describe the input and output:

Input — an array of real values and the number of valid entries in the array
Output — the average, minimum, and maximum values from the array

Step 3 is to work a simple example by hand. However, these operations are all ones that we have used in examples before, so we do not repeat the hand examples here.

Step 4 is to develop an algorithm. Since we have already developed algorithms for determining averages, minimums, and maximums, the steps for the subroutine are straightforward.

Decomposition

Determine xmin, xmax, xsum.
Compute xave.
Return.

Pseudocode

Stat(x,n,xave,xmin,xmax):
  sum ← x(1)
  xmin ← x(1)
  xmax ← x(1)
  For i = 2 to n do
    sum ← sum + x(i)
    If x(i) < xmin then xmin ← x(i)
    If x(i) > xmax then xmax ← x(i)
  xave ← sum/n
  Return

FORTRAN Subroutine

```
*------------------------------------------------------------------*
      SUBROUTINE STAT(X,N,XAVE,XMIN,XMAX)
*
*  This subroutine computes the average, minimum, and
*  maximum of a real array with N values.
*
      INTEGER N, I
      REAL X(N), XAVE, XMIN, XMAX, SUM
*
      SUM = X(1)
      XMIN = X(1)
      XMAX = X(1)
*
      DO 10 I=2,N
         SUM = SUM + X(I)
         IF (X(I).LT.XMIN) XMIN = X(I)
         IF (X(I).GT.XMAX) XMAX = X(I)
   10 CONTINUE
*
      XAVE = SUM/REAL(N)
*
      RETURN
      END
*------------------------------------------------------------------*
```

Step 5 is to test the subroutine. We use a program that reads a set of exam scores and then calls the subroutine STAT to compute some statistics from the scores.

FORTRAN Program

```
*------------------------------------------------------------------*
      PROGRAM SCORES
*
*  This program reads a set of test scores and then uses a
*  subroutine to determine the average, minimum, and maximum.
```

```
*
      INTEGER N, I
      REAL TESTS(100), AVE, MIN, MAX
*
      PRINT*, 'ENTER NUMBER OF TESTS (<101)'
      READ*, N
      PRINT*, 'ENTER TEST SCORES'
      READ*, (TESTS(I),I=1,N)
*
      CALL STAT(TESTS,N,AVE,MIN,MAX)
*
      PRINT 5, AVE
    5 FORMAT (1X,'AVERAGE TEST SCORE = ',F6.2)
      PRINT 10, MIN, MAX
   10 FORMAT (1X,'MINIMUM SCORE = ',F6.2,5X,'MAXIMUM SCORE = ',
   +          F6.2)
*
      END
*------------------------------------------------------------------*
                   STAT subroutine goes here
*------------------------------------------------------------------*
```

EXAMPLE 7-2   ## Sort Subroutine

In Chapter 5, we wrote a program that included the steps to sort a one-dimensional array. This operation is used so frequently that we will rewrite it in the form of a subroutine, using the bubble sort algorithm. To make it flexible, we use a variable in the argument list to specify the number of elements in the array. We suggest that you store this subroutine where it can be accessed easily.

SOLUTION

Step 1 is to state the problem clearly: Write a subroutine to sort an array of data.

Step 2 is to describe the input and output:

Input — an array of real values and the number of valid entries in the array
Output — an array of the reordered data

Step 3 is to work a simple example by hand. Since we worked a hand example in Chapter 5 for the bubble sort algorithm, you can refer to page 133 if you need to review it.

Step 4 is to develop an algorithm. Since we developed this algorithm in Chapter 5, we need only add the steps so that the solution is implemented as a subroutine.

FORTRAN Subroutine

```
*------------------------------------------------------------------*
      SUBROUTINE SORT(X,Y,N)
*
*  This subroutine sorts an array X into an array Y in
*  ascending order.  Both arrays have N values.
```

```
*
      INTEGER N, I, FIRST, LAST
      REAL X(N), Y(N), TEMP
      LOGICAL SORTED
*
      DO 10 I=1,N
         Y(I) = X(I)
   10 CONTINUE
*
      SORTED = .FALSE.
      FIRST = 1
      LAST = N - 1
   15 IF (.NOT.SORTED) THEN
         SORTED = .TRUE.
         DO 20 I=FIRST,LAST
            IF (Y(I).GT.Y(I+1)) THEN
               TEMP = Y(I)
               Y(I) = Y(I+1)
               Y(I+1) = TEMP
               SORTED = .FALSE.
            END IF
   20    CONTINUE
         LAST = LAST - 1
         GO TO 15
      END IF
*
      RETURN
      END
*------------------------------------------------------------------*
```

Step 5 is to test the subroutine. We use the following program that reads a list of data, uses the subroutine to sort them, and then prints the data in the new order.

FORTRAN Program

```
*------------------------------------------------------------------*
      PROGRAM LIST
*
* This program reads a list of real values, sorts them into
* ascending order using a subroutine, and then prints the
* reordered values.
*
      INTEGER N, I
      REAL X(25), Y(25)
*
      PRINT*, 'ENTER THE NUMBER OF VALUES TO SORT'
      PRINT*, '(MAXIMUM 25)'
      READ*, N
      PRINT*, 'ENTER THE VALUES'
      READ*, (X(I),I=1,N)
*
      CALL SORT(X,Y,N)
*
      PRINT*, 'VALUES IN ASCENDING ORDER:'
```

```
      DO 10 I=1,N
         PRINT 5, Y(I)
   5     FORMAT (1X,F5.2)
  10 CONTINUE
*
      END
*------------------------------------------------------------*
                   SORT subroutine goes here
*------------------------------------------------------------*
```

In the main program, we specified that the list of values to be sorted had a maximum of 25 values (because we dimensioned our arrays to 25 values). The subroutine itself does not have any maximum on the array size; it can sort any number of values as long as the values are part of an array that has been properly defined in the main program.

If you do not wish to use two arrays with the sort subroutine, you can use this CALL statement:

$$\text{CALL SORT(X,X,N)}$$

Remember, however, that the original order of the values is lost.

. . . .

Try It   Try this self-test to check your memory of some key points from Section 7-1. If you have any problems with these exercises, you should reread this section. The solutions are given at the end of this module.

Questions 1 and 2 refer to the following program and subroutine. The program generates an array of 10 numbers containing the integers 1 – 10. The subroutine modifies these numbers.

```
*------------------------------------------------------------*
      PROGRAM QUIZ
*
*  This program tests your understanding of subroutines.
*
      INTEGER K(10), I
*
      DO 10 I=1,10
         K(I) = I
  10 CONTINUE
*
      CALL MODIFY(K,10)
*
      PRINT*, 'NEW VALUES OF K ARE:'
      PRINT 15, (K(I),I=1,10)
  15 FORMAT (1X,5I5)
*
      END
*------------------------------------------------------------*
      SUBROUTINE MODIFY(K,N)
*
*  This subroutine modifies elements in K.
```

```
*
      INTEGER N, K(N), I
*
      DO 5 I=1,N
         K(I) = MOD(K(I),4)
    5 CONTINUE
*
      RETURN
      END
*----------------------------------------------------------------*
```

1. What is the program output?
2. What is the program output if the reference to the subroutine is replaced with the following statement?

```
CALL MODIFY(K,8)
```

7-2 Application   SIMULATION DATA

Electrical Engineering

A routine to generate random numbers is useful in many engineering and science applications. Most game programs use randomly generated numbers to make the program appear to have a mind of its own; it chooses different actions each time the game is played. Programs that simulate something, such as tosses of a coin or the number of people at a bank window, also use random number generators.

In this application, we use a random number generator to simulate (or model) noise, such as static, that might occur in a piece of instrumentation. The routine generates numbers between 0.0 and 1.0. The numbers are uniformly distributed across the interval between 0.0 and 1.0, which means that we are just as likely to get 0.4455 as we are to get 0.0090. This random number generator requires an argument that is a seed to the computation. When we give the routine a different seed, it returns a different random number. To generate a sequence of random numbers, we initialize the seed once and we do not modify it again. The random number generator modifies the seed itself from one call of the routine to the next and thus must be implemented as a subroutine since it has two outputs. The details of the following random number generator[1] are beyond the scope of this text. It essentially causes the computer to compute integers that are too large to store. The portion of the number that can be stored is a random sequence that is used to determine the random number.

FORTRAN Subroutine

```
*----------------------------------------------------------------*
      SUBROUTINE RANDOM(SEED,RANDX)
*
*  This subroutine generates a random number between 0.0 and
*  1.0. An integer seed is used to initialize the sequence.
```

---

[1]S. D. Stearns, "A Portable Random Number Generator for Use in Signal Processing," Sandia National Laboratory Technical Report (1981).

```
*
      INTEGER SEED
      REAL RANDX
*
      SEED = 2045*SEED + 1
      SEED = SEED - (SEED/1048576)*1048576
      RANDX = REAL(SEED + 1)/1048577.0
*
      RETURN
      END
*-----------------------------------------------------------------*
```

The following main program allows you to enter a seed; the program then prints the first 10 random numbers generated with that seed. Try the program with different seeds and observe that you get different numbers. An example output is shown for a seed of 12357.

FORTRAN Program

```
*-----------------------------------------------------------------*
      PROGRAM TEST
*
*  This program tests the random number generator.
*
      INTEGER I, SEED
      REAL X
*
      PRINT*, 'ENTER A POSITIVE INTEGER SEED VALUE'
      READ*, SEED
      PRINT*, 'RANDOM NUMBERS:'
      DO 10 I=1,10
         CALL RANDOM(SEED,X)
         PRINT 5, X
    5    FORMAT (1X,F8.6)
   10 CONTINUE
*
      END
*-----------------------------------------------------------------*
                        RANDOM subroutine goes here
*-----------------------------------------------------------------*
```

Sample Output

```
ENTER A POSITIVE INTEGER SEED VALUE
12357
RANDOM NUMBERS:
0.099414
0.299419
0.310731
0.442812
0.548521
0.725532
0.712078
0.199705
0.395030
0.834445
```

We are now ready to look at the application problem. We want to develop a program that generates a data file containing a sine wave plus noise. The program will use both the intrinsic sine function and the subroutine to generate random numbers. The data file will contain values of this signal

$$f(t) = 2 \sin(2\pi t) + \text{noise}$$

for $t = 0.0, 0.01, \ldots, 1.00$. Each value of the sine function is added to a random number produced by the random number generator. Since the sine wave can vary from -2 to $+2$ and the random number generator can vary from 0 to 1, we expect this experimental signal to vary from -2 to 3. The signal should be stored in a data file called SIGNAL, where each line of the data file contains the value of t and the corresponding signal value.

1. Problem Statement

Generate a data file that contains samples of the function $2 \sin(2\pi t)$, with uniform random noise between 0.0 and 1.0 added to it. The data points are to be evaluated with $t = 0.0, 0.01, \ldots, 1.00$.

2. Input/Output Description

Input—the seed to start the random number generator
Output—a data file named SIGNAL

3. Hand Example

The discussion of the random number generator illustrated the first 10 random numbers generated with seed 12357. Using these random numbers, the first 10 data points of the file SIGNAL are

$$
\begin{aligned}
f(0.00) &= 2 \cdot \sin(2\pi \cdot 0.00) + 0.099414 = 0.0994144 \\
f(0.01) &= 2 \cdot \sin(2\pi \cdot 0.01) + 0.299419 = 0.4250000 \\
f(0.02) &= 2 \cdot \sin(2\pi \cdot 0.02) + 0.310731 = 0.5613975 \\
f(0.03) &= 2 \cdot \sin(2\pi \cdot 0.03) + 0.442812 = 0.8175746 \\
f(0.04) &= 2 \cdot \sin(2\pi \cdot 0.04) + 0.548521 = 1.0459008 \\
f(0.05) &= 2 \cdot \sin(2\pi \cdot 0.05) + 0.725532 = 1.3435660 \\
f(0.06) &= 2 \cdot \sin(2\pi \cdot 0.06) + 0.712078 = 1.4483271 \\
f(0.07) &= 2 \cdot \sin(2\pi \cdot 0.07) + 0.199705 = 1.0512636 \\
f(0.08) &= 2 \cdot \sin(2\pi \cdot 0.08) + 0.395030 = 1.3585374 \\
f(0.09) &= 2 \cdot \sin(2\pi \cdot 0.09) + 0.834445 = 1.9060986
\end{aligned}
$$

4. Algorithm Development

The only input is the random seed. The program then generates the signal values and writes them in a data file. We do not need arrays because each data value is needed only once.

Decomposition

Read random number seed.
Generate data values and write them to the file.

Flowchart

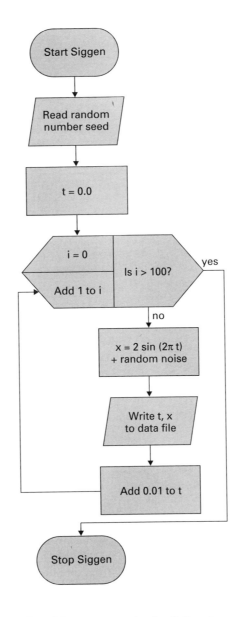

The structure chart for this program is the following:

FORTRAN Program

```
*------------------------------------------------------------------*
      PROGRAM SIGGEN
*
* This program generates a data file that contains a
* signal composed of a sine wave plus random noise. The
* user enters a seed to initialize the random noise signal.
*
      INTEGER SEED, I
      REAL PI, T, NOISE, X
*
      PARAMETER (PI=3.141593)
*
      OPEN (UNIT=15,FILE='SIGNAL',STATUS='NEW')
*
      PRINT*, 'ENTER A POSITIVE INTEGER SEED VALUE'
      READ*, SEED
*
      T = 0.0
      DO 10 I=1,101
         CALL RANDOM(SEED,NOISE)
         X = 2.0*SIN(2.0*PI*T) + NOISE
         WRITE (15,*) T, X
         T = T + 0.01
   10 CONTINUE
*
      END
*------------------------------------------------------------------*
              RANDOM subroutine goes here
*------------------------------------------------------------------*
```

5. Testing

Use the random number generator seed that we used in the hand example. Check the first 10 values of the data signal in the data file; if these match, we can be certain that our random number generator is working. It is difficult to test a program with random values thoroughly because we cannot expect (and should not get) the same values if we change the seed for the random number generator. However, the trend of the data should be the same. For example, in this problem we should be able to see the sine wave in the data even though we use different random numbers for the noise. A plotting routine is helpful here. (In Chapter 8, problem 12 discusses a simple plot routine that can be used to plot data on your terminal screen.) Some plots of the data from program SIGGEN are shown. The first plot shows the sine wave with no noise; the next two plots are plots of the data file using different random seeds. These plots were generated using MATLAB.

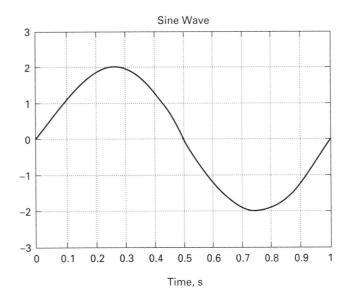

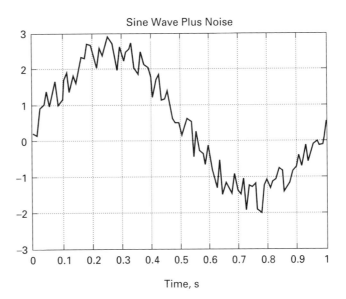

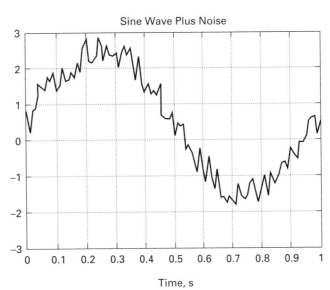

7-3 **Common Blocks**

As you modularize your programs, you will find that the argument lists can become lengthy as you pass more and more data to functions and subroutines. FORTRAN allows you to set up a block of memory that is accessible, or common, to the main program and to all its subprograms—a *common block*. The variables in this block of memory do not have to be passed through argument lists and cannot be used as arguments in subprograms.

FORTRAN allows two types of common blocks: *blank common* and *named common*. Blank common is set up with the following specification statement:

COMMON *variable list*

Each subprogram that uses data in this common block must also contain a COMMON statement. Although the data names do not have to match in every subprogram, the type and order of the names are important. Consider these COMMON statements from a main program and a subprogram, respectively:

```
PROGRAM TEST1
INTEGER J
REAL A, B, X, Y
COMMON A, J, B
.
.
.
CALL ANSWR(X,Y)
.
.
.
END
SUBROUTINE ANSWR(X,Y)
INTEGER KTOT
REAL TEMP, SUM, X, Y
COMMON TEMP, KTOT, SUM
.
.
.
END
```

In these statements, the main program communicates with the subroutine using two arguments, X and Y. In addition, the main program and the subroutine share a memory area. A and TEMP represent the same memory location, J and KTOT represent the same memory location, and B and SUM represent the same memory location, as shown in the following diagram:

Common Memory Area

A	57.63	TEMP
J	−25	KTOT
B	0.007	SUM

Named common, also referred to as *labeled common,* is established if the list of variable names in the COMMON statement is preceded by a name enclosed in slashes:

COMMON */name/variable list*

The purpose of establishing different blocks of common with unique names is to allow subprograms to refer to the named common block with which they wish to share data without listing all the other variables in the other common blocks.

For example, one group of variables may be used in a set of computations, and another group of variables may be used in generating data files that will be plotted. The first group of variables could comprise one common area named CALC, and the other group of variables could comprise another common area named GRAPH. Both named COMMON statements appear in the main program; then individual subprograms include either or both named COMMON statements depending on the variables they need to access. The argument lists for the individual subprograms include other input and output variables that are not in the common areas. A variable cannot appear in more than one block of named common; thus, a variable could not appear in both the CALC and GRAPH common blocks. If the variable is needed in the computations and the graphics, it should be passed through the argument list of the subprograms or in a third named common block.

Arrays may be included in the COMMON statement. The COMMON statement can be used to define the storage for these arrays as shown:

```
COMMON X(50), INCOME(4)
```

An array size can be specified in the COMMON statement, as shown in the previous statement, or it can be specified in an explicit type statement, as in

```
INTEGER SUM
REAL TIME(25), DIST(25)
COMMON SUM, TIME, DIST
```

The size of an array cannot be specified in both a type statement and a COMMON statement.

Variables in blank common cannot be initialized with DATA statements. They can be initialized with READ statements or assignment statements. Variables in named common can be initialized with READ statements, assignment statements, or with a special subprogram called a *BLOCK DATA subprogram.* This subprogram is nonexecutable and serves only to assign initial values to variables in a named common block. An example of a BLOCK DATA subprogram to initialize two named common blocks is

```
BLOCK DATA
COMMON /EXPER1/ TEMP(100)
COMMON /EXPER2/ TIME, DIST, VEL(10)
DATA TEMP, TIME, DIST, VEL /100*0.0, 50.5, 0.5, 10*0.0/
END
```

Both of the COMMON statements must be in the main program. Either or both of the statements also appear in any subprogram that uses variables in these common blocks of memory.

Unless extremely large amounts of data must be passed to subprograms, the use of common blocks is discouraged because it weakens the independence of modules; the module can be affected not only by its arguments but also by variables in common. If you do use common blocks, pay attention to the order of the variables on all COMMON statements. Also look for omitted variables; if a variable is left out of a COMMON statement in a subprogram, incorrect values are used for all variables following it in the COMMON statement. These errors can cause problems that are difficult to locate.

SUMMARY

The subroutine is a subprogram that allows us to return multiple values to the main program, such as when we want to sort an array, or to return no values to the main program, such as when we want to print an error message. Functions and other subroutines can also call a subroutine, in addition to the main program. The subroutine, along with the function, allows us to structure our programs by breaking them into modules that can be written and tested independently of each other. Not only do programs become simpler, but also the same module can be used in other programs without retesting. This chapter developed a sort routine.

Key Words

blank common	labeled common
BLOCK DATA subprogram	named common
common block	subroutine

Problems

We begin our problem set with modifications to programs developed earlier in this chapter. Give the decomposition, refined pseudocode or flowchart, and FORTRAN solution for each problem.

Problem 1 modifies the program SIGGEN given in Section 7-2, which generated a data file composed of a sine wave plus random noise.

1. Modify the signal generating program so that it reads the number of data points that are to be generated from the terminal, instead of generating 101 points.

For problems 2–4, assume that TIME, a one-dimensional array of 100 real values, has already been filled with data.

2. Write a subroutine to determine the maximum value in the array TIME and to subtract that value from each value in the array.
3. Write a subroutine to determine the average of the values in the array TIME. Add 1.0 to all values above the average; subtract 1.0 from all values below the average.
4. Write a subroutine to return a logical value of true if all the values in the array TIME are zero; otherwise, the value returned should be false.

For problems 5–6, assume that Z, a two-dimensional array with five rows and four columns of real values, has already been filled with data.

5. Write a subroutine that fills an array W (the same size as Z) with a value based on the corresponding element in Z, where

$$
\begin{aligned}
W(I,J) &= 0.0 && \text{if } Z(I,J) = 0 \\
W(I,J) &= -1.0 && \text{if } Z(I,J) < 0 \\
W(I,J) &= 1.0 && \text{if } Z(I,J) > 0
\end{aligned}
$$

6. Write a subroutine that fills an array W (the same size as Z) with numbers such that each element of W is the corresponding value of Z rounded up to the next multiple of 10. Thus, 10.76 rounds to 20.0, 18.7 rounds to 20.0, 0.05 rounds to 10.0, 10.0 rounds to 10.0, and -5.76 rounds to 0.0.

Develop these programs and subroutines. Use the five-step design process.

7. Write a subroutine that computes the average, the variance, and the standard deviation of an array X of 100 data values. Use the following formulas:

$$
\text{Average:} \qquad \overline{X} = \frac{\displaystyle\sum_{i=1}^{100} X_i}{100}
$$

$$
\text{Variance:} \qquad \sigma^2 = \frac{\displaystyle\sum_{i=1}^{100} (\overline{X} - X_i)^2}{99}
$$

$$
\text{Standard deviation:} \quad \sigma = \sqrt{\sigma^2}
$$

8. Rewrite the subroutine in problem 7 so that it computes the average, variance, and standard deviation for an array with 500 values. Assume that N contains the number of actual values in the array. The denominator of the expression for the average should be N; the denominator of the expression for the variance should then be $N - 1$.

9. Write a subroutine that receives a two-dimensional real array X with 50 rows and 2 columns and returns the same array with the data reordered. Sort the data such that the values in the second column are in ascending order. The values in the first column should correspond to the values in the second column. That is, the same values should be on a row together in both the original order and the new order, but the ordering of the rows may change.

10. Write a subroutine that receives a two-dimensional real array X with 50 rows and 4 columns and returns the same array with the data reordered. Sort the data such that the values in each row are in ascending order.

11. Write a subroutine that receives a two-dimensional real array X with 50 rows and 2 columns and returns the same array with the data reordered. Sort the data such that the values in each column are in descending order.

12. Write a subroutine called BIAS that is called with the following statement:

CALL BIAS(X,Y,N)

where X is an input array of 200 real values, N is an integer that specifies how many of the values represent actual data values, and Y is an output array the same size as X whose values should be the values of X with the minimum value of the X array subtracted from each one. For example, if

$$X = \boxed{10 \mid 2 \mid 36 \mid 8}$$

then

$$Y = \boxed{8 \mid 0 \mid 34 \mid 6}$$

Thus, the minimum value of Y is always zero. (This operation is referred to as removing the bias in X or adjusting for bias in X.)

13. Write a subroutine that receives a two-dimensional real array X with 50 rows and 4 columns and returns the same array with the data reordered. An additional argument J is used to select a column that is to be sorted in descending order. The other values are not to be changed.

14. Write a subroutine that receives a two-dimensional real array X with 50 rows and 4 columns and returns the same array with the data reordered. An additional argument J is used to select a column that is to be sorted in descending order. The other values are to be changed so that values on the same row stay together in the new reordering.

8 Additional Data Types

Genetic Engineering Genetic engineering begins with a gene that produces a valuable substance such as the human growth hormone. Enzymes are used to dissolve bonds to the neighboring genes, thus separating the valuable gene out of the DNA. This gene is then inserted into another organism, such as a bacterium, that will multiply itself along with the foreign gene. One step in discovering a valuable gene is identifying the sequence of amino acids in the protein that it produces. A protein sequencer is a sophisticated piece of equipment that can determine the order of amino acids making up a chainlike protein molecule, thus uncovering the identity of the gene that made it. Although there are only 20 different amino acids, protein molecules have hundreds of amino acids linked in a specific order.

This chapter presents three data types: character, double precision, and complex. The character data type allows us to read and analyze character data such as chemical formulas. With double-precision data, we can process numeric data more precisely than we could using previously discussed data types. With complex data, we can represent data as numbers that have a real portion and an imaginary portion. Although we do not use these data types routinely, they are special features of FORTRAN that help make it a powerful language for engineering and scientific applications.

8-1 CHARACTER DATA

You have probably already learned that computers internally use a binary language that is composed of 0s and 1s. Integers and real numbers are converted to binary numbers when they are used in a computer. If you study computer hardware or computer architecture, you learn how to convert values such as 56 and -13.25 to binary numbers; to use FORTRAN, however, it is not necessary to learn this conversion.

Characters also must be converted into binary form to be used in the computer; they are converted to *binary strings,* which are also sequences of 0s and 1s. Several codes convert character information to binary strings, but most computers use *EBCDIC* (Extended Binary Coded Decimal Interchange Code) or *ASCII* (American Standard Code for Information Interchange). In these codes, each character is represented by a binary string. Table 8–1 contains a few characters and their EBCDIC and ASCII equivalents.

You do not need to use binary codes to use the characters in your FORTRAN programs. However, you must be aware that the computer stores characters differently than the numbers used in arithmetic computations; that is, the integer number 5 and the character 5 are not stored in the same way. Thus, it is not possible to use arithmetic operations with character data even if the characters represent numbers.

We often refer to character data as *character strings* because we usually refer to groups or lists of characters that go together. For example, a chemical formula is usually given one variable name instead of a variable name for each character in the formula. For example, we might use the name WATER for the string 'H2O'. We can have character string constants that always represent the same information. Character string variables have names and represent character strings that may remain constant or may change. Generally, these character string constants and variables contain characters from the *FORTRAN character set,* which is composed of the 26 alphabetic letters, the 10 numeric digits, a blank, and the following 12 symbols:

$$+ \quad - \quad * \quad / \quad = \quad (\quad) \quad , \quad . \quad ' \quad \$ \quad :$$

If other symbols are used, a program may not execute the same way on one computer as it does on another.

Character constants are always enclosed in apostrophes. These apostrophes are not counted when determining the length or number of characters in a constant. If two consecutive apostrophes (not a double quotation mark) are encountered within a character constant, they represent a single apostrophe. For example, the character constant for the word LET'S is 'LET''S'.

Table 8-1   Binary Character Codes

Character	ASCII	EBCDIC
A	1000001	11000001
H	1001000	11001000
Y	1011001	11101000
3	0110011	11110011
+	0101011	01001110
$	0100100	01011011
=	0111101	01111110

The following list gives several examples of character constants and their corresponding lengths:

Constant	**Length**
'SENSOR 23'	9 characters
'TIME AND DISTANCE'	17 characters
' $ AMT.'	7 characters
' '	2 characters
'08:40-13:25'	11 characters
''''''	2 characters

A character string variable is defined with a specification statement whose general form is

CHARACTER*n *variable list*

where *n* represents the number of characters in each character string. For instance, the statement

    CHARACTER*8 CODE, NAME

identifies CODE and NAME as variables containing eight characters each. Unlike numeric variable names, there is no significance to the first letter of a character variable's name. A variation of the CHARACTER statement allows you to specify character strings of different lengths in the same statement, as shown:

CHARACTER TITLE*10, STATE*2

An array that contains character strings can be defined using either of the following statements:

CHARACTER*4 NAME(50)
CHARACTER NAME(50)*4

The preceding specifications reserve memory for 50 elements in the array NAME, where each element contains four characters. A reference to

NAME(18) references a character string with four characters that is the 18th element of the array.

Character strings can also be used as arguments in subprograms. The character string must be specified in a CHARACTER statement in both the main program and the subprogram. A subprogram can specify a character string without giving a specific length, as in

```
CHARACTER*(*) STRING
```

This technique is similar to specifying the length of an array with an integer variable, as in

```
INTEGER SSN(N)
```

It is also possible to define in a subprogram an array of N variables, each of which contains a character string. To make the character string more flexible, its length does not have to be specified in the CHARACTER statement. We use the following statement in the subprogram to accomplish this flexibility:

```
CHARACTER*(*) NAME(N)
```

Later, this section will present special operations and intrinsic functions for character strings. First, we look at how to use character strings in input and output statements.

Character I/O

When a character string is used in a list-directed output statement, the entire character string is printed. Blanks are automatically inserted around the character string to separate it from other output on the same line. When a character string variable is used in a list-directed input statement, the corresponding data value must be enclosed in apostrophes. If the character string within the apostrophes is longer than the defined length of the character string variable, any extra characters on the right are ignored; if the character string within the apostrophes is shorter than the length of the character string variable, the extra positions to the right are automatically filled with blanks. To print a character string in a formatted output statement, use A as the corresponding format specification to print the entire string.

EXAMPLE 8-1

Character I/O

Write a complete FORTRAN program to read an item description from the terminal and then print the description. Assume that the length of the description is no more than 20 characters.

SOLUTION

We start with the decomposition.

Decomposition

Read description.
Print description.

The following refinement indicates the conversation with the user:

Pseudocode

    Output: Print message to user to enter description
            Read description
            Print description

Translating these steps into FORTRAN results in the following program:

FORTRAN Program

```
*---------------------------------------------------------------*
      PROGRAM OUTPUT
*
*   This program reads and prints an item description.
*
      CHARACTER*20 ITEM
*
      PRINT*, 'ENTER ITEM DESCRIPTION IN APOSTROPHES'
      READ*, ITEM
*
      PRINT 5, ITEM
    5 FORMAT (1X,'ITEM DESCRIPTION IS ',A)
*
      END
*---------------------------------------------------------------*
```

A sample interaction with this program is

```
ENTER ITEM DESCRIPTION IN APOSTROPHES
'COMPUTER MODEM'
ITEM DESCRIPTION IS COMPUTER MODEM
```

Note that the data entered was not 20 characters long. In this example, the padding of blanks on the end is not noticeable with the output; however, if the output FORMAT had been

```
FORMAT(1X,A,' IS THE ITEM DESCRIPTION')
```

the interaction would have the following appearance:

```
ENTER ITEM DESCRIPTION IN APOSTROPHES
'COMPUTER MODEM'
COMPUTER MODEM        IS THE ITEM DESCRIPTION
```

Another interaction that could come from the original program is

```
ENTER ITEM DESCRIPTION IN APOSTROPHES
'DIGITAL OSCILLOSCOPE WITH MEMORY'
ITEM DESCRIPTION IS DIGITAL OSCILLOSCOPE
```

In this case, the name exceeded the maximum number of characters specified for the description, so part of the data was lost. To avoid this situation, carefully choose the length of your character variables based on the maximum length you expect. You can also tell the program user what length you are expecting. In this example, you could use the following pair of PRINT statements:

```
PRINT*, 'ENTER ITEM DESCRIPTION IN APOSTROPHES'
PRINT*, '(MAXIMUM OF 20 CHARACTERS)'
```

Character Operations

Although character strings cannot be used in arithmetic computations, we can assign values to character strings, compare two character strings, extract a substring of a character string, and combine two character strings into one longer character string.

Assign Values   Values can be assigned to character variables with the assignment statement and a character constant. The following statements initialize a character string array RANK with the five abbreviations for freshman, sophomore, junior, senior, and graduate:

```
CHARACTER*2 RANK(5)
RANK(1) = 'FR'
RANK(2) = 'SO'
RANK(3) = 'JR'
RANK(4) = 'SR'
RANK(5) = 'GR'
```

If a character constant in an assignment statement is shorter in length than the character variable, blanks are added to the right of the constant. If the following statement was executed, RANK(1) would contain the letter *F* followed by a blank:

```
RANK(1) = 'F'
```

If a character constant in an assignment statement is longer than the character variable, the excess characters on the right are ignored. Thus, the following statement,

```
RANK(1) = 'FRESHMAN'
```

would store the letters *FR* in the character array element RANK(1). These examples emphasize the importance of using character strings that are the same length as the variables used to store them; otherwise, the statements would be misleading.

**Table 8-2   Partial Collating Sequences
for Characters**

ASCII

ᵦ " # $ % & ( ) * + , - . /

0 1 2 3 4 5 6 7 8 9

: ; = ? @

A B C D E F G H I J K L M N O P Q R S T U V W X Y Z

EBCDIC

ᵦ . (+ & $ *) ; - / , % ? : # @ = "

A B C D E F G H I J K L M N O P Q R S T U V W X Y Z

0 1 2 3 4 5 6 7 8 9

One character string variable can also be used to initialize another character string variable, as shown:

```
CHARACTER*4 GRADE1, GRADE2
GRADE1 = 'GOOD'
GRADE2 = GRADE1
```

Both variables, GRADE1 and GRADE2, contain the character string 'GOOD'.

Character strings can be initialized with DATA statements. The preceding examples can be performed with a DATA statement, as shown:

```
CHARACTER RANK(5)*2, GRADE1*4, GRADE2*4
DATA RANK, GRADE1, GRADE2 /'FR', 'SO', 'JR', 'SR',
+                                        'GR', 2*'GOOD'/
```

Compare Values   An IF statement can be used to compare character strings. Assuming that the variable DEPT and the array CH are character strings, the following are valid statements:

```
IF (DEPT.EQ.'EECE') KT = KT + 1

IF (CH(I).GT.CH(I+1)) THEN
   CALL SWITCH(I,CH)
   CALL PRINT(CH)
END IF
```

To evaluate a logical expression using character strings, first look at the length of the two strings. If one string is shorter than the other, add blanks to the right of the shorter string so that you can proceed with the evaluation using strings of equal length.

The comparison of two character strings of the same length is made from left to right, one character at a time. Two strings must have exactly the same characters in the same order to be equal.

A *collating sequence* lists characters from the lowest to the highest value. Partial collating sequences for EBCDIC and ASCII are given in Table 8-2. Although the ordering is not exactly the same, some similarities include

1. Capital letters are in order from *A* to *Z*.

2. Digits are in order from 0 to 9.
3. Capital letters and digits do not overlap; either digits precede letters or letters precede digits.
4. The blank character is less than any letter or number. When necessary for clarity, we use $_b$ to represent a blank.

The following is a list of several pairs of character strings, along with their correct relationships:

$$\begin{aligned}
\text{'A1'} &< \text{'A2'} \\
\text{'JOHN'} &< \text{'JOHNSTON'} \\
\text{'176'} &< \text{'177'} \\
\text{'THREE'} &< \text{'TWO'} \\
\text{'\$'} &< \text{'DOLLAR'}
\end{aligned}$$

If character strings contain only letters, their ordering from low to high is alphabetical, which is also called a *lexicographic ordering*.

Extract Substrings A *substring* of a character string is any string that represents a subset of the original string and maintains the original order. The following list contains all substrings of the string 'FORTRAN':

'F'	'FO'	'FOR'	'FORT'	'FORTR'	'FORTRA'	'FORTRAN'
'O'	'OR'	'ORT'	'ORTR'	'ORTRA'	'ORTRAN'	
'R'	'RT'	'RTR'	'RTRA'	'RTRAN'		
'T'	'TR'	'TRA'	'TRAN'			
'R'	'RA'	'RAN'				
'A'	'AN'					
'N'						

Substrings are referenced by using the name of the character string, followed by two integer expressions in parentheses, separated by a colon. The first expression in parentheses gives the position in the original string of the beginning of the substring; the second expression gives the position of the end of the substring. If the string 'FORTRAN' is stored in a variable LANG, some of its substring references are as shown:

Reference	Substring
LANG(1:1)	'F'
LANG(1:7)	'FORTRAN'
LANG(2:3)	'OR'
LANG(7:7)	'N'

If the first expression in parentheses is omitted, the substring begins at the beginning of the string; thus, LANG(:4) refers to the substring 'FORT'. If the second expression in parentheses is omitted, the substring ends at the end of the string; thus, LANG(5:) refers to the substring 'RAN'.

The substring operation cannot operate correctly if the beginning and ending positions are not integers, are negative, or contain values greater than the number of characters in the substring. The ending position must also be greater than or equal to the beginning position of the substring.

The substring operator is a powerful tool, as the next example illustrates.

EXAMPLE 8-2   ## Character Count

A string of 50 characters contains encoded information. The number of occurrences of the letter *S* represents a special piece of information. Write a loop that counts the number of occurrences of the letter *S*.

SOLUTION

The loop index is used with the substring operator to allow us to test each character in the string:

```
CHARACTER*50 CODE
INTEGER COUNT, I
DATA COUNT /0/
    .
    .
    .
    DO 20 I=1,50
        IF (CODE(I:I).EQ.'S') COUNT = COUNT + 1
 20 CONTINUE
```

. . . .

A reference to a substring can be used anywhere that a string can be used. For instance, if LANG contains the character string 'FORTRAN', the following statement changes the value of LANG to 'FORMATS':

$$LANG(4:7) = 'MATS'$$

If LANG contains 'FORMATS', the following statement changes the value of LANG to 'FORMATT':

$$LANG(7:7) = LANG(6:6)$$

When modifying a substring of a character string with a substring of the same character string, the substrings must not overlap — that is, do not use LANG(2:4) to replace LANG(3:5). Also, recall that if a substring is being moved into a smaller string, only as many characters as are needed to replace the smaller string are moved from left to right; if the substring is being moved into a larger string, the extra positions on the right are filled with blanks.

Combine Strings   *Concatenation* is the operation of combining two or more character strings into one character string. It is indicated by two slashes between the character strings to be combined. The following expression concatenates the constants 'WORK' and 'ED' into one string constant 'WORKED':

$$'WORK'//'ED'$$

The next statement combines the contents of three character string variables MO, DA, and YR into one character string and then moves the combined string into a variable called DATE:

$$\text{DATE} = \text{MO}//\text{DA}//\text{YR}$$

If MO = '05', DA = '15', and YR = '86', then DATE = '051586'. Because concatenation represents an operation, it cannot appear on the left-hand side of an equal sign.

Character Intrinsic Functions

A number of intrinsic functions are designed for use with character strings:

INDEX locates specific substrings within a given character string.
LEN determines the length of a string and is used primarily in subroutines and functions that have character string arguments.
CHAR and ICHAR determine the position of a character in the collating sequence of the computer.
LGE, LGT, LLE, and LLT allow comparisons to be made based on the ASCII collating sequence, regardless of the collating sequence of the computer.

Index   The INDEX function has two arguments, both of which are character strings. The function returns an integer value giving the position in the first string of the second string. Thus, if STRGA contains the phrase 'TO BE OR NOT TO BE', INDEX(STRGA,'BE') returns the value 4, which points to the first occurrence of the string 'BE'. To find the second occurrence of the string, we can use the following statements:

```
CHARACTER*18 STRGA
.
.
.
K = INDEX(STRGA,'BE')
J = INDEX(STRGA(K+1:),'BE') + K
```

After executing these statements, K would contain the value 4, and J would contain the value 17. (Note that we had to add K to the second reference of INDEX because the second use referred to the substring 'E OR NOT TO BE', and thus the second INDEX reference returns a value of 13, not 17.) The value of INDEX(STRGA,'AND') would be 0 because the second string 'AND' does not occur in the first string STRGA.

Len   The input to the function LEN is a character string; the output is an integer that contains the length of the character string. This function is useful in a subprogram that accepts character strings of any length but needs the actual length within the subprogram. The statement in the subprogram that allows a character string to be defined without specifying its length is

```
CHARACTER*(*) A, B, STRGA
```

This form can be used only in subprograms. Example 8-3 uses both the LEN function and a variable string length in a subprogram.

EXAMPLE 8-3

Frequency of Blanks

Write a function subprogram that accepts a character string and returns a count of the number of blanks in the string.

SOLUTION

To make this function flexible, we write it so that it can be used with any size character string.

```
*------------------------------------------------------------------*
      INTEGER FUNCTION BLANKS(X)
*
*  This function counts the number of blanks in a
*  character string X.
*
      INTEGER I
      CHARACTER*(*) X
*
      BLANKS = 0
      DO 10 I=1,LEN(X)
         IF (X(I:I).EQ.' ') BLANKS = BLANKS + 1
   10 CONTINUE
*
      RETURN
      END
*------------------------------------------------------------------*
```

· ·

Character strings may also be used in user-written subroutines. In Example 8-4, we write a subroutine that combines input character strings into an output character string.

EXAMPLE 8-4

Name Editing

Write a subroutine that receives 3 character strings, FIRST, MIDDLE, and LAST, each containing 15 characters. The output of the subroutine is a character string 35 characters long that contains the first name followed by 1 blank, the middle initial followed by a period and 1 blank, and the last name. Assume that FIRST, MIDDLE, and LAST have no leading blanks and no embedded blanks. Thus, if

```
             FIRST = 'JOSEPH          '
             MIDDLE = 'CHARLES         '
             LAST = 'LAWTON          '
```

then the edited name would be:

```
      JOSEPH C. LAWTON
```

SOLUTION

The solution to this problem is simplified by the use of the substring operation that allows us to look at individual characters and the INDEX function
that is used to find the end of the first name. We move to NAME the characters in FIRST, then a blank, the middle initial, a period, another blank, and the
last name. As you go through the solution, observe the use of the concatenation operation. Also, note that the move of the first name fills the rest of the
character string NAME with blanks because FIRST is shorter than the field to
which it is moved:

```
*----------------------------------------------------------------*
      SUBROUTINE EDIT(FIRST,MIDDLE,LAST,NAME)
*
*  This subroutine edits a name to the form
*  first, middle initial, last.
*
      INTEGER L
      CHARACTER*15 FIRST, MIDDLE, LAST
      CHARACTER*35 NAME
*
*  MOVE FIRST NAME
*
      NAME = FIRST
*
*  MOVE MIDDLE INITIAL
*
      L = INDEX(NAME,' ')
      NAME(L:L+3) = ' '//MIDDLE(1:1)//'. '
*
*  MOVE LAST NAME
*
      NAME(L+4:) = LAST
*
      RETURN
      END
*----------------------------------------------------------------*
```

. . . .

CHAR, ICHAR   These functions refer to the collating sequence used in the
computer. If a computer has 50 characters in its collating sequence, these
characters are numbered from 0 to 49. For example, assume that the letter A
corresponds to position number 12. The function CHAR uses an integer
argument that specifies the position of a desired character in the collating
sequence, and the function returns the character in the specified position.
The following statements print the character A:

```
        N = 12
        PRINT*, CHAR(N)
```

The ICHAR function is the inverse of the CHAR function. The argument to the
ICHAR function is a character variable that contains one character. The
function returns an integer that gives the position of the character in the

collating sequence. Thus, the output of the following statements is the number 12:

```
CHARACTER*1 INFO
INFO = 'A'
PRINT*, ICHAR(INFO)
```

Because different computers have different collating sequences, these functions can be used to determine the position of certain characters in the collating sequence.

EXAMPLE 8-5

Collating Sequence

Print each character in the FORTRAN character set along with its position in the collating sequence on your computer. The FORTRAN character set is given on page 208.

SOLUTION

Note the use of the substring operator in initializing the character set:

```
*--------------------------------------------------------------------*
      PROGRAM SEQNCE
*
*  This program prints the position in the collating sequence
*  of each FORTRAN character.
*
      CHARACTER*49 SET
*
      SET(1:26) = 'ABCDEFGHIJKLMNOPQRSTUVWXYZ'
      SET(27:36) = '0123456789'
      SET(37:49) = ' +-*/=(),.''$:'
*
      DO 10 I=1,49
         PRINT*, SET(I:I), ICHAR(SET(I:I))
   10 CONTINUE
*
      END
*--------------------------------------------------------------------*
```

Why did we put two apostrophes in the assignment for SET(37:49)? Remember that two apostrophes are converted into a single apostrophe when they are in a literal. If you have several computers available, run this program on each of them to see if they all use the same collating sequence for the FORTRAN character set.

. . . .

LGE, LGT, LLE, LLT This set of functions allows you to compare character strings based on the ASCII collating sequence. These functions become useful if a program is going to be used on a number of different computers and is using character comparisons or character sorts. The functions represent a logical value, true or false. Each function has two character string arguments, STRG1 and STRG2. The function reference LGE(STRG1,STRG2) is true if STRG1 is lexically greater than or equal to STRG2; thus, if STRG1 comes after STRG2 in an alphabetical sort, this function reference is true. Remember, these functions are based on an ASCII collating sequence regardless of the sequence being used on the computer. The functions LGT, LLE, and LLT perform comparisons lexically greater than, lexically less than or equal to, and lexically less than.

Try It

Try this self-test to check your memory of some key points from Section 8-1. If you have any problems with the exercises, you should reread this section. The solutions are given at the end of this module.

For problems 1–10, give the substring referred to in each reference. Assume that a character string of length 35 called TITLE has been initialized with the statements

```
CHARACTER*35 TITLE
TITLE = 'TEN TOP ENGINEERING ACHIEVEMENTS'
```

1. TITLE(1:20)
2. TITLE(1:8)
3. TITLE(9:19)
4. TITLE(21:21)
5. TITLE(9:)
6. TITLE(:8)
7. TITLE(:)
8. TITLE(1:4)//TITLE(21:)
9. TITLE(5:8)//TITLE(1:3)
10. TITLE(9:16)//TITLE(32:32)//''' '//TITLE(21:)

In problems 11–16, WORD is a character string of length 6. What is stored in WORD after each of the following statements?

11. WORD = 'LASER'
12. WORD = 'FIBER OPTICS'
13. WORD = 'CAD'//'CAM'
14. WORD = ''''''''
15. WORD = ' 12.48'
16. WORD ='GENETIC ENGINEERING'

8-2 Application   PROTEIN MOLECULAR WEIGHTS

Genetic Engineering
Genetic engineering begins with a gene that produces a valuable substance such as the human growth hormone that was discussed in the chapter-opening application. Enzymes are used to dissolve bonds to the

Table 8-3   Amino Acids

Amino Acid	Reference	Molecular Weight
Glycine	Gly	75
Alanine	Ala	89
Valine	Val	117
Leucine	Leu	131
Isoleucine	Ile	131
Serine	Ser	105
Threonine	Thr	119
Tyrosine	Tyr	181
Phenylalanine	Phe	165
Tryptophan	Trp	203
Aspartic	Asp	132
Glutamic	Glu	146
Lysine	Lys	147
Arginine	Arg	175
Histidine	His	156
Cysteine	Cys	121
Methionine	Met	149
Asparagine	Asn	132
Glutamine	Gln	146
Proline	Pro	116

neighboring genes, thus separating the valuable gene out of the DNA. This gene is then inserted into another organism, such as a bacterium, that will multiply itself along with the foreign gene.

One step in discovering a valuable gene is identifying the sequence of amino acids in the protein that it produces. A protein sequencer is a sophisticated piece of equipment that can determine the order of amino acids making up a chainlike protein molecule, thus uncovering the identity of the gene that made it. Although there are only 20 different amino acids, protein molecules have hundreds of amino acids linked in a specific order.

In this problem, we assume that the sequence of amino acids in a protein molecule has been identified and that we want to compute the molecular weight of the protein molecule. Table 8–3 lists the amino acids, their three-letter reference, and their molecular weights. A data file named AMINO contains 20 lines of data; each line of the data file contains a three-letter reference (enclosed in quotation marks) and the corresponding molecular weight.

Assume that another data file contains the protein molecule characterizations in amino acids. The first line in the file contains the number of protein molecules in the file, and each following line contains a character string in quotes that contains the amino acid sequence. The maximum number of amino acids in a character string is 50. The program should determine and print the corresponding molecular weight for each protein. Print an error message if an incorrect protein string is detected.

1. Problem Statement

Write a program that will determine the molecular weight of a group of protein molecules containing only amino acids.

2. Input/Output Description

Input—a data file containing the information on amino acids and a different data file containing the protein molecules

Output—the molecular weights of the protein molecules

3. Hand Example

Suppose that the protein molecule is the following:

LysGluMetAspSerGlu

The corresponding molecular weights for the amino acids are

147,146,149,132,105,146

Therefore, the protein molecular weight is 825.

4. Algorithm Development

We start the algorithm development with the decomposition.

Decomposition

| Read and store amino acid data in arrays. |
| Read protein molecules and compute and print the molecular weight. |

Pseudocode

Weight: Read amino strings and weights into arrays
         Read the number of proteins, n
         For k = 1 to n do
              sum ← 0
              Read protein string
              For each amino acid in the protein string
                   Add corresponding weight to the sum
              Print protein string and sum

The step to add the weight for each amino acid to the molecular weight sum requires comparing each amino acid string to the reference strings and then selecting the corresponding weight. We will implement this step in a subroutine in order to keep the main program easy to read. Since there is only a small number of amino acids to test, we use a DO loop. If the amino acid is found, the corresponding weight will be moved to a variable named WEIGHT; if the amino acid is not found, a value of zero will be returned in WEIGHT.

```
MWT (ref,mw,string,weight):
    weight ← 0
    For k=1 to 20 do
        if ref(k) = string then
            weight ← mw(k)
    Return
```

There are several things to note about the following program. For example, we used the INDEX function to determine the first blank in the protein string. This allowed us to determine the number of amino acids and also to print only the nonblank characters; otherwise, PROTN would require 150 output characters. Also, note that if we find an error in an amino acid, we continue evaluating the protein string. This allows us to catch all the errors at once, instead of catching only one error per protein at a time.

FORTRAN Program

```
*------------------------------------------------------------*
      PROGRAM WEIGHT
*
* This program reads character strings containing amino acids
* from large protein molecules and computes molecular weights.
*
      INTEGER K, MW(20), N, J, BLNK, NCHAR, AMNUM, SUM, START,
     +       AMWT
      CHARACTER*3 REF(20)
      CHARACTER*150 PROTN
      LOGICAL ERROR
*
      OPEN (UNIT=9,FILE='AMINO',STATUS='OLD')
      DO 10 K=1,20
         READ(9,*) REF(K), MW(K)
   10 CONTINUE
*
      OPEN (UNIT=10,FILE='PROTEIN',STATUS='OLD')
      READ(10,*) N
*
      DO 30 J=1,N
*
         READ(10,*) PROTN
         BLNK = INDEX(PROTN,' ')
         NCHAR = BLNK - 1
         AMNUM = NCHAR/3
         SUM = 0
*
```

```
                IF (MOD(NCHAR,3).NE.0) THEN
*
                   PRINT*, 'LENGTH ERROR IN PROTEIN ', J
                   PRINT*, PROTN(:BLNK)
                   PRINT*
*
                ELSE
*
                   ERROR = .FALSE.
                   DO 20 K=1,AMNUM
                      START = (K-1)*3 + 1
                      CALL MWT(REF,MW,PROTN(START:START+2),AMWT)
                      IF (AMWT.NE.0) THEN
                         SUM = SUM + AMWT
                      ELSE
                         PRINT*, 'ERROR IN AMINO ', K, ' PROTEIN ', J
                         ERROR = .TRUE.
                      END IF
 20                CONTINUE
                   IF (.NOT.ERROR) THEN
                      PRINT 15, PROTN(:BLNK), SUM
 15                   FORMAT (1X,'PROTEIN:  ',A/
     +                        1X,'MOLECULAR WEIGHT:',I9/)
                   ELSE
                      PRINT*, PROTN(:BLNK)
                      PRINT*
                   END IF
*
                END IF
*
 30 CONTINUE
*
       END
*----------------------------------------------------------------*
       SUBROUTINE MWT(REF,MW,STRING,WEIGHT)
*
*  This subroutine receives arrays containing the character
*  references and molecular weights for amino acids. It uses
*  these arrays to determine if an input string is an amino acid
*  and, if so, returns the molecular weight of the amino acid;
*  otherwise zero is returned.
*
       INTEGER MW(20), WEIGHT, K
       CHARACTER*3 REF(20), STRING
*
       WEIGHT = 0
       DO 10 K=1,20
          IF (STRING.EQ.REF(K)) WEIGHT = MW(K)
 10 CONTINUE
*
       RETURN
       END
*----------------------------------------------------------------*
```

◢ 5. Testing

The following PROTEIN file was used for testing:

```
5
'GlyIle'
'AspHisProGln'
'ThrTbrSerTrpLys'
'AlaValLeuValMet'
'LysGluMetAspSerGlu'
```

The output for the test file was the following:

```
PROTEIN:  GlyIle
MOLECULAR WEIGHT:        206

PROTEIN:  AspHisProGln
MOLECULAR WEIGHT:        550

ERROR IN AMINO    2    PROTEIN    3
ThrTbrSerTrpLys

PROTEIN:  AlaValLeuValMet
MOLECULAR WEIGHT:        603

PROTEIN:  LysGluMetAspSerGlu
MOLECULAR WEIGHT:        825
```

8-3 Double-Precision Data

Double-precision variables are necessary any time we want to keep more significant digits than are stored in real variables. Assume that a real variable can store 7 significant digits; this means that the real variable will keep 7 digits of accuracy, beginning with the first nonzero digit, in addition to remembering where the decimal point goes. A *double-precision value* can store more digits, with the exact number of digits dependent on the computer. For this discussion, assume that double-precision variables store 14 digits. The following table compares values that can be stored in real values (also called single-precision values) and in double-precision values.

Value to be Stored	Single Precision	Double Precision
37.6892718	37.68927	37.689271800000
− 1.60003	− 1.600030	− 1.6000300000000
820000000487.	820000000000.	820000000487.00
18268296.300405079	18268290.	18268296.300405

Note that we are doubling the precision of our values, but we are not doubling the range of numbers that can be stored. The same range of numbers applies to both single- and double-precision values, but double-precision values store those numbers with more digits of precision.

Many science and engineering applications use double-precision values to increase accuracy. For example, the study of solar systems, galaxies, and stars requires storing immense distances with as much precision as possible. Even economic models often need double precision. A model for predicting the gross national debt must handle numbers exceeding $1 trillion. With single precision, you cannot store the values with an accuracy to the nearest dollar. With double precision, amounts up to $100 trillion can be used and still have significant digits for all dollar amounts.

A double-precision constant is written in an exponential form, with a D in place of the E. Some examples of double-precision constants are

```
0.378926542D+04
1.4762D-02
0.25D+00
```

Always use the exponential form with the letter D for double-precision constants, even if the constant uses seven or fewer digits of accuracy; otherwise, you may lose some accuracy because a fractional value that can be expressed exactly in decimal notation may not be expressed exactly in binary notation.

Double-precision variables are specified with a specification statement whose general form is

```
DOUBLE PRECISION variable list
```

A double-precision array is specified as

```
DOUBLE PRECISION DTEMP(50)
```

Double-Precision I/O

Double-precision variables can be used in list-directed output in the same manner that we list real values. The only distinction is that more digits of accuracy can be stored in a double-precision value; therefore, more digits of accuracy can be written from a double-precision value.

In formatted input and output, double-precision values may be referenced with the F or E format specifications. Another specification, Dw.d, may also be used. Dw.d functions essentially like the E specification, but the D emphasizes that it is being used with a double-precision value. In output, the value in this exponential form is printed with a D instead of an E. Thus, if the following statements were executed,

```
      DOUBLE PRECISION DX
         .
         .
         .
      DX = 1.66587514521D+00
      PRINT 10, DX
   10 FORMAT (1X,'DX = ',D17.10)
```

the output would be

$$DX = 0.1665875145D+01$$

EXAMPLE 8-6 ## Solar Distances

Assume that a character array has been filled with the names of planets, moons, and other celestial bodies. A corresponding array has been filled with the distances of these objects from the sun in millions of miles. Both arrays have been defined to hold 200 values, and an integer N specifies how many elements are actually stored in the arrays. The array NAME is an array of character strings of length 20, and the array DIST is a double-precision array. Give the statements to print these names and distances.

SOLUTION

Before printing the data in the arrays, we print a heading for the data. Next, a loop is executed N times and is used to print the object name and its corresponding distance from the sun in millions of miles. We use a D format for the output because the set of distances may cover a large range of values. The CHARACTER statement and the DOUBLE PRECISION statement are included with the statements to print the data:

```
      CHARACTER*20 NAME(200)
      DOUBLE PRECISION DIST(200)
        .
        .
        .
      PRINT*, 'SOLAR OBJECTS AND DISTANCES FROM THE SUN'
      PRINT*, '                    (MILLIONS OF MILES)'
      PRINT*
      DO 10 I=1,N
         PRINT 5, NAME(I), DIST(I)
    5    FORMAT (1X,A,2X,D15.8)
   10 CONTINUE
```

A sample output from these statements is

```
SOLAR OBJECTS AND DISTANCES FROM THE SUN
                    (MILLIONS OF MILES)

JUPITER              0.43863717D+03
MARS                 0.14151751D+03
SATURN               0.88674065D+03
VENUS                0.67235696D+02
PLUTO                0.36662718D+04
URANUS               0.17834237D+04
MERCURY              0.35979176D+02
NEPTUNE              0.27944448D+04
EARTH                0.92961739D+02
```

Double-Precision Operations

When an arithmetic operation is performed with two double-precision values, the result is double precision. If an operation involves a double-precision value and a single-precision value or an integer, the result is a double-precision result. In such a mixed-mode operation, do not assume that the other value is converted to double precision; instead, think of the other value as being extended in length with zeros. To illustrate, the first two assignment statements that follow yield exactly the same values; the third assignment statement, however, adds a double-precision constant to DX and yields the most accurate result of the three statements:

```
DOUBLE PRECISION DX, DY1, DY2, DY3
   .
   .
   .
DY1 = DX + 0.3872
DY2 = DX + 0.3872000000000
DY3 = DX + 0.3872D+00
```

The most accurate way to obtain a constant that cannot be written in a fixed number of decimal places is to perform a double-precision operation that yields the desired value. For instance, to obtain the double-precision constant one-third, use the following expression:

```
1.0D+00/3.0D+00
```

Using double-precision values increases the precision of our results, but there is a price for this additional precision: The execution time for computations is longer and more memory is required.

Double-Precision Intrinsic Functions

If a double-precision argument is used in a generic function, the function value is also double precision. Many of the common intrinsic functions for real numbers can be converted to double-precision functions by preceding the function name with the letter *D*. For instance, DSQRT, DABS, DMOD, DSIN, DEXP, DLOG, and DLOG10 all require double-precision arguments and yield double-precision values. Since the generic function and double-precision function perform the same operation with double-precision arguments, we recommend using the generic function name for consistency. Double-precision functions can also be used to compute constants with double-precision accuracy. For instance, the following statements compute π with double-precision accuracy:

```
DOUBLE PRECISION DPI
   .
   .
   .
DPI = 4.0D+00*ATAN(1.0D+00)
```

(Recall that $\pi/4$ is equal to the arctangent of 1.0.)

Although Appendix A contains a complete list of the functions that relate to double-precision values, two functions, DBLE and DPROD, are specifically designed for use with double-precision variables. DBLE converts a REAL argument to a double-precision value by adding zeros. DPROD has two real arguments and returns the double-precision product of the two arguments.

Try It

Try this self-test to check your memory of some key points from Section 8-3. If you have any problems with the exercises, you should reread this section. The solutions are given at the end of this module.

In problems 1–6, show how to represent these constants as double-precision constants.

1. 0.75

2. 1.3

3. 1/9

4. 5/6

5. –10.5

6. 3.0

In problems 7–9, show the output of the following PRINT statements if DX = 0.00786924379, where DX is a double-precision variable.

```
7.   PRINT 4, DX
     4 FORMAT (1X,D14.5)
8.   PRINT 5, DX
     5 FORMAT (1X,D12.3)
9.   PRINT 6, DX
     6 FORMAT (1X,F12.7)
```

8-4 COMPLEX DATA

Complex numbers are needed to solve many problems in science and engineering, particularly in physics and electrical engineering. Therefore, FORTRAN includes a special data type for complex variables and constants. (Recall that complex numbers have the form $a + bi$, where i is $\sqrt{-1}$ and a and b are real numbers. Thus, the real part of the number is represented by a, and the imaginary part of the number is represented by b.) These *complex values* are stored as an ordered pair of real values that represents the real and the imaginary portions of the value.

A complex constant is specified by two real constants separated by a comma and enclosed in parentheses. The first constant represents the real part of the complex value; the second constant represents the imaginary part of the complex value. Thus, the complex constant $3.0 + 1.5i$ is written in FORTRAN as the complex constant (3.0, 1.5).

Complex variables are specified with a specification statement, whose general form is

COMPLEX *variable list*

A complex array is specified as

```
COMPLEX CX(100)
```

Complex I/O

A complex value in list-directed output is printed as two real values separated by a comma and enclosed in parentheses. Two real values are read for each complex value in a list-directed input statement.

For formatted output, a complex value is printed with two real specifications. The real part of the complex value is printed before the imaginary portion. It is good practice to enclose the two parts printed in parentheses and separate them by a comma or print them in the $a + bi$ form. Both forms are illustrated in the following statements:

```
COMPLEX CX, CY
.
.
.
CX = (1.5, 4.0)
CY = (0.0, 2.4)
PRINT 5, CX, CY
5  FORMAT (1X,'(',F4.1,',',F4.1,')'/1X,F4.1,' + ',F4.1,' i')
```

The output from the PRINT statement is

```
( 1.5, 4.0)
0.0 +  2.4 i
```

Complex Operations

When an arithmetic operation is performed between two complex values, the result is also a complex value. In an expression containing a complex value and a real or integer value, the real or integer value is converted to a complex value whose imaginary part is zero. Expressions containing both complex values and double-precision values are not allowed.

The rules for complex arithmetic are not as familiar as those for integers or real values. Table 8–4 lists the results of basic operations on two complex numbers C_1 and C_2, where $C_1 = a_1 + b_1 i$ and $C_2 = a_2 + b_2 i$.

Complex Intrinsic Functions

If a complex value is used in one of the generic functions, such as SQRT, ABS, SIN, COS, EXP, or LOG, the function value is also complex. The functions CSQRT, CABS, CSIN, CCOS, CEXP, and CLOG are all intrinsic functions with

Table 8-4   Complex Arithmetic Operations

Operation	Result
$C_1 + C_2$	$(a_1 + a_2) + i\,(b_1 + b_2)$
$C_1 - C_2$	$(a_1 - a_2) + i\,(b_1 - b_2)$
$C_1 \cdot C_2$	$(a_1 a_2 - b_1 b_2) + i\,(a_1 b_2 + a_2 b_1)$
$\dfrac{C_1}{C_2}$	$\left(\dfrac{a_1 a_2 + b_1 b_2}{a_2{}^2 + b_2{}^2}\right) + i\left(\dfrac{a_2 b_1 - b_2 a_1}{a_2{}^2 + b_2{}^2}\right)$
$|C_1|$	$\sqrt{a_1{}^2 + b_1{}^2}$
e^{C_1}	$e^{a_1}\cos b_1 + i\,e^{a_1}\sin b_1$
$\cos C_1$	$1 - \dfrac{C_1{}^3}{3!} + \dfrac{C_1{}^5}{5!} - \dfrac{C_1{}^7}{7!} + \cdots$

(Note that $C_1 = a_1 + ib_1$ and $C_2 = a_2 + ib_2$.)

complex arguments. These function names begin with the letter C to emphasize that they are complex functions.

Although Appendix A contains a complete list of the functions that relate to complex values, four functions, REAL, AIMAG, CONJG, and CMPLX, are specifically designed for use with complex variables: REAL yields the real part of its complex argument; AIMAG yields the imaginary part of its complex argument; CONJG converts a complex number to its conjugate, where the conjugate of $(a + bi)$ is $(a - bi)$; COMPLX converts two real arguments, a and b, into a complex value $(a + bi)$. Note that, while (2.0,1.0) is equal to the complex constant $2.0 + 1.0i$, we must use the expression CMPLX(A,B) to specify the complex variable $A + Bi$; the expression (A,B) by itself does not represent a complex variable in FORTRAN 77.

EXAMPLE 8-7

Quadratic Formula

The roots of a quadratic equation with real coefficients may be complex. Give the statements to compute and print the two roots of a quadratic equation, given the coefficients A, B, and C as shown:

$$AX^2 + BX + C = 0$$

$$X_1 = \frac{-B + \sqrt{B^2 - 4AC}}{2A} \qquad X_2 = \frac{-B - \sqrt{B^2 - 4AC}}{2A}$$

SOLUTION

```
COMPLEX DISCR, ROOT1, ROOT2
  .
  .
  .
DISCR = CMPLX(B*B - 4.0*A*C,0.0)
ROOT1 = (-B + SQRT(DISCR))/(2.0*A)
ROOT2 = (-B - SQRT(DISCR))/(2.0*A)
```

```
     PRINT*, 'ROOTS TO THE QUADRATIC EQUATION ARE:'
     PRINT 5, ROOT1, ROOT2
   5 FORMAT (1X,F5.2,' + ',F5.2,' i',4X,F5.2,' + ',F5.2,' i')
```

Two sets of sample output are shown, one in which both roots are real and one in which both roots are complex:

```
ROOTS TO THE QUADRATIC EQUATION ARE:
 1.56 +  0.00 i    2.00 +  0.00 i

ROOTS TO THE QUADRATIC EQUATION ARE:
 2.36 +  2.45 i    2.36 + -2.45 i
```

. . . .

Try It

Try this self-test to check your memory of some key points from Section 8-4. If you have any problems with the exercises, you should reread this section. The solutions are given at the end of this module.

In problems 1–6, compute the value stored in CX if CY = $2 - i$ and CZ = $-1 + 2i$. Assume that CX, CY, and CZ are complex variables.

1. CX = CY + 2*CZ
2. CX = CZ - CY
3. CX = CONJG(CY)
4. CX = REAL(CZ) + AIMAG(CY)
5. CX = CMPLX(3.5,-1.5)
6. CX = ABS(CY + CZ)

In problems 7 and 8, show the output of the following PRINT statements. Assume that CX = $2 - i$.

7. PRINT 5, CX
    5 FORMAT (1X,F5.1,'+',F5.1,' i')
8. PRINT 6, CX
    6 FORMAT (1X,'(',F6.2,',',F6.2,')')

Summary

With the data types presented in this chapter, we now have a number of choices for defining our variables and constants. For numeric data, we can select integers, real values, or double-precision values. We have character data for information that is not going to be used in numeric computations. For special applications, we have complex variables. In addition to being able to choose the proper data type for our data, we also have special intrinsic functions and operations for simplifying our work with the data.

Key Words

ASCII	collating sequence
binary string	complex value
character string	concatenation

double-precision value   lexicographic order
EBCDIC   substring
FORTRAN character set

Problems

This problem set begins with modifications to programs developed earlier in this chapter. Give the decomposition, refined pseudocode or flowchart, and FORTRAN solution for each problem.

Problems 1 – 5 modify the program WEIGHT given in Section 8-2, which computes the molecular weight of a large protein molecule.

1. Modify the molecular weight program so that it includes a subroutine that is used to print the contents of the AMINO file at the beginning of the program.
2. Modify the molecular weight program so that it prints a final line with the maximum and minimum protein molecular weights.
3. Modify the molecular weight program so that it prints a final line with the average protein molecular weight.
4. Modify the molecular weight program so that it prints the number of amino acids in each protein molecule.
5. Modify the molecular weight program so that it prints the maximum and minimum number of amino acids in the protein molecules.

In problems 6 – 16, develop programs and modules using the five-step design process.

6. Write a complete program that reads a data file ADDR containing 50 names and addresses. The first line for each person contains the first name (10 characters), the middle name (6 characters), and the last name (21 characters). The second line contains the address (25 characters), the city (10 characters), the state abbreviation (2 characters), and the zip code (5 characters). Each character string is enclosed in quotes in the file. Print the information in the following label form:

```
First Initial. Middle Initial. Last Name
Address
City, State Zip
```

Skip four lines between labels. The city should not contain any blanks before the comma that follows it. A typical label might be

```
J. D. Doe
117 Main St.
Taos, NM 87166
```

For simplicity, assume there are no embedded blanks in the individual data values. For example, San Jose would be entered as SanJose and printed in the same manner.

7. Write a complete program to read a double-precision value from the terminal. Compute the sine of the value using the following series:

$$\sin X = X - \frac{X^3}{3!} + \frac{X^5}{5!} - \frac{X^7}{7!} + \cdots$$

Continue using terms until the absolute value of a term is less than 1.0D − 09. Print the computed sine and the value obtained from the sine function for comparison.

8. A palindrome is a word or piece of text that is spelled the same forward and backward. The word 'RADAR' is an example of a palindrome, but ' RADAR' is not a palindrome because of the unmatched blank. 'ABLE ELBA' is another palindrome. Write a logical function PALIND that receives a character variable X of length 20. The function should be true if the character array is a palindrome; otherwise, it should be false.

9. Write a subroutine ALPHA that receives an array of 50 letters. The subroutine should alphabetize the list of letters.   26

10. Modify the subroutine in problem 9 to remove duplicate letters and to add blanks at the end of the array for the letters removed.

11. Write a subroutine that receives a piece of text called PROSE that contains 200 characters. The subprogram should print the text in lines of 30 characters each. Do not split words between two lines. Do not print any lines that are completely blank.

12. Write a subroutine that receives an array of *N* real values (maximum value of *N* is 200) and generates a printer plot. Use an output line of 101 characters. Scale the line from the minimum value to the maximum value. The first line of output should be 101 periods representing the *Y*-axis. All the following lines should contain a period in the column representing $X = 0$ (if this point is included in the graph), and the letter *X* in the position of each data point, as shown in the following diagram. (In this diagram, the minimum and the maximum values have the same absolute value.)

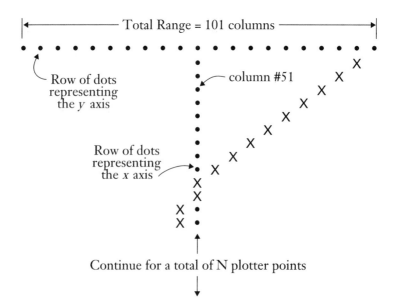

Continue for a total of N plotter points

13. Write a complete program that reads and stores the following two-dimensional array of characters:

```
ATIDEB
LENGTH
ECPLOT
DDUEFS
OUTPUT
CGDAER
HIRXJI
KATIMN
BHPARG
```

Now read the 11 strings that follow and find the same strings in the preceding array. Print the positions of characters of these hidden words that may appear forward, backward, up, down, or diagonally. For instance, the word EDIT is located in positions (1,5), (1,4), (1,3), and (1,2).

Hidden Words

PLOT	STRING
CODE	TEXT
EDIT	READ
LENGTH	GRAPH
INPUT	BAR
OUTPUT	

14. Write a subroutine that sorts a complex array of N values into a descending order based on the absolute value of the complex values. Assume that the subroutine is called with the statement

```
CALL ORDER(CDATA,N)
```

where CDATA is an array of N complex values.

15. Write a subroutine ENCODE that receives a character string KEY containing 26 characters and a character string MESSGE of an unspecified size. The subroutine should encode MESSGE, using a substitution code where the first letter in KEY is substituted for the letter A in MESSGE, the second letter in KEY is substituted for the letter B in MESSGE, and so on. Thus, if KEY contained the character string

```
'YXAZKLMBJOCFDVSWTREGHNIPUQ'
```

and MESSGE contained the character string

```
'MEET AT AIRPORT SATURDAY'
```

then the encoded character string would be

```
'DKKG YG YJRWSRG EYGHRZYU'
```

The encoded character string should be stored in a character string called SECRET. The subroutine is called with the following statement:

```
                                CALL ENCODE(KEY,MESSGE,SECRET)
```

16. Write a subroutine DECODE that receives a character string KEY containing 26 characters and a character string SECRET of an unspecified size. The subroutine should decode SECRET, which has been encoded with a substitution key where the first letter in KEY was substituted for the letter *A*, the second letter in KEY was substituted for the letter *B*, and so on. Blanks were not changed in the coding process. Thus, if KEY contained the character string

```
                        'YXAZKLMBJOCFDVSWTREGHNIPUQ'
```

and SECRET contained the character string

```
                          'DKKG YG YJRWSRG EYGHRZYU'
```

then the decoded character string would be

```
                          'MEET AT AIRPORT SATURDAY'
```

The decoded character string should be stored in a character string called MESSGE. The subroutine is called with the statement

```
                          CALL DECODE(KEY,SECRET,MESSGE)
```

Appendix A

FORTRAN 77 INTRINSIC FUNCTIONS

In the following table of intrinsic functions, the names of the arguments specify their type, as indicated below:

Argument		Type
X	→	real
CHX	→	character
DX	→	double precision
CX	→	complex, a + bi
LX	→	logical
IX	→	integer
GX	→	generic

Function type, the second column of the table of intrinsic functions, specifies the type of value returned by the function.

Generic function names are printed in bold. Any type argument that is applicable can be used with generic functions, and the function value returned is generally the same type as the input arguments, except for type conversion functions such as REAL and INT.

Function Name	Function Type	Definition
SQRT(GX)	Same as GX	$\sqrt{GX}$
DSQRT(DX)	Double precision	$\sqrt{DX}$
CSQRT(CX)	Complex	$\sqrt{CX}$
ABS(GX)	Same as GX	$|GX|$
IABS(IX)	Integer	$|IX|$
DABS(DX)	Double precision	$|DX|$
CABS(CX)	Complex	$|CX|$
EXP(GX)	Same as GX	e^{GX}
DEXP(DX)	Double precision	e^{DX}
CEXP(CX)	Complex	e^{CX}

Function Name	Function Type	Definition
LOG(GX)	Same as GX	$\log_e GX$
ALOG(X)	Real	$\log_e X$
DLOG(DX)	Double precision	$\log_e DX$
CLOG(CX)	Complex	$\log_e CX$
LOG10(GX)	Same as GX	$\log_{10} GX$
ALOG10(X)	Real	$\log_{10} X$
DLOG10(DX)	Double precision	$\log_{10} DX$
REAL(GX)	Real	Convert GX to real value
FLOAT(IX)	Real	Convert IX to real value
SNGL(DX)	Real	Convert DX to single precision
ANINT(GX)	Same as GX	Round to nearest whole number
DNINT(DX)	Double precision	Round to nearest whole number
NINT(GX)	Integer	Round to nearest integer
IDNINT(DX)	Integer	Round to nearest integer
AINT(GX)	Same as GX	Truncate GX to whole number
DINT(DX)	Double precision	Truncate DX to whole number
INT(GX)	Integer	Truncate GX to an integer
IFIX(X)	Integer	Truncate X to an integer
IDINT(DX)	Integer	Truncate DX to an integer
SIGN(GX, GY)	Same as GX, GY	Transfer sign of GY to $|GX|$
ISIGN(IX, IY)	Integer	Transfer sign of IY to $|IX|$
DSIGN(DX, DY)	Double precision	Transfer sign of DY to $|DX|$
MOD(GX, GY)	Same as GX, GY	Remainder from GX/GY
AMOD(X, Y)	Real	Remainder from X/Y
DMOD(DX, DY)	Double precision	Remainder from DX/DY
DIM(GX, GY)	Same as GX, GY	GX − (minimum of GX and GY)
IDIM(IX, IY)	Integer	IX − (minimum of IX and IY)
DDIM(DX, DY)	Double precision	DX − (minimum of DX and DY)
MAX(GX,GY, . . .)	Same as GX, GY, . . .	Maximum of (GX,GY, . . .)
MAX0(IX,IY, . . .)	Integer	Maximum of (IX,IY, . . .)
AMAX1(X,Y, . . .)	Real	Maximum of (X,Y, . . .)
DMAX1(DX,DY, . . .)	Double precision	Maximum of (DX,DY, . . .)
AMAX0(IX,IY, . . .)	Real	Maximum of (IX,IY, . . .)
MAX1(X,Y, . . .)	Integer	Maximum of (X,Y, . . .)
MIN(GX,GY, . . .)	Same as GX, GY, . . .	Minimum of (GX,GY, . . .)
MIN0(IX,IY, . . .)	Integer	Minimum of (IX,IY, . . .)
AMIN1(X,Y, . . .)	Real	Minimum of (X,Y, . . .)
DMIN1(DX,DY, . . .)	Double precision	Minimum of (DX,DY, . . .)
AMIN0(IX,IY, . . .)	Real	Minimum of (IX,IY, . . .)
MIN1(X,Y, . . .)	Integer	Minimum of (X,Y, . . .)
SIN(GX)	Same as GX	Sine of GX, assumes radians
DSIN(DX)	Double precision	Sine of DX, assumes radians
CSIN(CX)	Complex	Sine of CX
COS(GX)	Same as GX	Cosine of GX, assumes radians
DCOS(DX)	Double precision	Cosine of DX, assumes radians
CCOS(CX)	Complex	Cosine of CX
TAN(GX)	Same as GX	Tangent of GX, assumes radians
DTAN(DX)	Double precision	Tangent of DX, assumes radians
ASIN(GX)	Same as GX	Arcsine of GX
DASIN(DX)	Double precision	Arcsine of DX
ACOS(GX)	Same as GX	Arccosine of GX
DACOS(DX)	Double precision	Arccosine of DX

Function Name	Function Type	Definition
ATAN(GX)	Same as GX	Arctangent of GX
DATAN(DX)	Double precision	Arctangent of DX
ATAN2(GX,GY)	Same as GX, GY	Arctangent of GX/GY
DATAN2(DX,DY)	Double precision	Arctangent of DX/DY
SINH(GX)	Same as GX	Hyperbolic sine of GX
DSINH(DX)	Double precision	Hyperbolic sine of DX
COSH(GX)	Same as GX	Hyperbolic cosine of GX
DCOSH(DX)	Double precision	Hyperbolic cosine of DX
TANH(GX)	Same as GX	Hyperbolic tangent of GX
DTANH(DX)	Double precision	Hyperbolic tangent of DX
DPROD(X,Y)	Double precision	Product of X and Y
DBLE(GX)	Double precision	Convert GX to double precision
CMPLX(GX)	Complex	$GX + 0i$
CMPLX(GX,GY)	Complex	$GX + GYi$
AIMAG(CX)	Real	Imaginary part of CX
REAL(CX)	Real	Real part of CX
CONJG(CX)	Complex	Conjugate of CX, $a - bi$
LEN(CHX)	Integer	Length of character string CHX
INDEX(CHX,CHY)	Integer	Position of substring CHY in string CHX
CHAR(IX)	Character	Character in the IXth position of collating sequence
ICHAR(CHX)	Integer	Position of the character CHX in the collating sequence
LGE(CHX,CHY)	Logical	Value of (CHX is lexically greater than or equal to CHY)
LGT(CHX,CHY)	Logical	Value of (CHX is lexically greater than CHY)
LLE(CHX,CHY)	Logical	Value of (CHX is lexically less than or equal to CHY)
LLT(CHX,CHY)	Logical	Value of (CHX is lexically less than CHY)

Appendix B

AN INTRODUCTION TO FORTRAN 90

This appendix reviews some of the new features of Fortran 90. The first section presents the expanded character set and the improved source format of the statements. The next section covers some of the new features relative to data objects. Since arrays are such an important data structure in many engineering problem solutions, one section focuses on a discussion of some of the enhancements relative to arrays. Another section covers the expansion of the DO statement. The final section discusses pointers — a new data object that allows us to build linked data structures.

B-1 EXPANDED CHARACTER SET AND SOURCE FORMAT

The Fortran character set includes the 26 alphabetic letters, the 10 numeric digits, and the underscore character (_). The set of special characters available has been expanded from 13 characters in FORTRAN 77 to the set of 21 characters shown in Table B-1. These special characters are used for operations and to separate components in a Fortran statement. A character literal can now be enclosed in either single or double quotation marks.

In addition to the standard character set and the set of special characters, Fortran 90 allows compilers to include other character sets. These character sets can be used for special purposes and include symbols for mathematics, chemistry, and music. To incorporate additional languages, character sets such as Kanji (Chinese), Greek, Cyrillic (Russian and other Slavic languages), Hindi, Magyar (Hungarian), and Nihongo (Japanese) can be included.

The characters that can be included in a name have been expanded from just letters and digits to include the underscore character. In addition, the maximum length has been expanded from 6 characters to 31 characters. If the compiler permits lowercase letters, they are equivalent to the uppercase letters except within character string data. Fortran 90 allows the use of alternative relational operators, as shown in the Table B-2.

Table B-1   Fortran 90 Special Characters

Character	Character Name	Character	Character Name
	Blank	:	Colon
=	Equals	!	Exclamation point
+	Plus	"	Quotation mark
−	Minus	%	Percent
*	Asterisk	&	Ampersand
/	Slash	;	Semicolon
(	Left parenthesis	<	Less than
)	Right parenthesis	>	Greater than
,	Comma	?	Question mark
.	Decimal point or period	$	Currency symbol
'	Apostrophe		

Fortran 90 allows two types of source forms: free form and fixed form. A program must be in one or the other form; it cannot mix the forms in the same program. Fixed source form is the form accepted by FORTRAN 77. In addition, comments can also be indicated by the character !; a comment initiated by the character ! can be on a line by itself or can follow a Fortran statement. In free source form, a line can contain no more than 132 characters, and there are no restrictions on where a statement may appear in the line. The character & is used at the end of the current line to indicate that the current statement is to be continued on the next line. Comments are indicated by the character ! and can appear on lines by themselves or can follow a Fortran statement. In both fixed and free format, a semicolon can be used to separate multiple statements on the same line.

B-2   INTRINSIC AND DERIVED DATA TYPES

Fortran 90 contains five intrinsic data types that are defined as part of the language: integer, real, complex, character, and logical. Derived data types can be defined from the intrinsic data types. For example, we might want to define a structure that contains a chemical name and its molecular weight. In FORTRAN 77, the only structure is an array, and array elements have to be the same type; therefore we could not include a chemical name and its molecular weight in the same array. In Fortran 90 programs, we can define structures using combinations of the default data types. An example of a defined data type is the following:

```
TYPE CHEMICAL
    CHARACTER (LEN30) :: NAME
    REAL :: WEIGHT
END TYPE CHEMICAL
```

To define a variable with this type, we use the following statement:

```
TYPE (CHEMICAL) :: AMINO_ACID
```

Table B-2   Relational Operators

Relational Operation	FORTRAN 77	Fortran 90 Addition
Equal	.EQ.	==
Not equal	.NE.	/=
Less than	.LT.	<
Less than or equal	.LE.	<=
Greater than	.GT.	>
Greater than or equal	.GE.	>=

To reference a component within this structure, the name of the structure is followed by a percent sign, followed by the component name. Therefore, to refer to the weight of the amino acid, we use AMINO__ACID % WEIGHT.

B-3   ARRAY ENHANCEMENTS

In Fortran 90, array names without subscripts can be used in expressions and assignments. These statements are interpreted element by element. A number of new intrinsic functions that apply specifically to arrays were added to Fortran 90. We list a number of these functions, along with a brief description in Table B-3.

In Fortran 90, we can initially define an array without specifying the size of the array. This is very useful when the size of the array is computed within the program or is read from a data file. In addition to allocating the memory needed for an array during the execution of a program, we can also release or deallocate the memory when we no longer need the array. The statements to allocate or deallocate memory are the following:

ALLOCATE *(array name (size))*

DEALLOCATE *(array name)*

B-4   CONTROL STATEMENTS

Fortran 90 controls the order in which statements within a program are executed using three kinds of control constructs: the IF construct, the CASE construct, and the DO construct. The IF construct is basically the same control structure that is included in FORTRAN 77; the CASE construct is a new control structure; the DO construct is an expanded version of the DO loop. Each of these constructs can contain statement blocks, which are groups of statements treated as a unit. One form of the DO construct is:

DO WHILE *(logical expression)*
            block of statements
END DO

Table B-3   Intrinsic Functions Involving Arrays

Function	Value Computed
DOT__PRODUCT(A, B)	Dot product of vectors A and B
MATMUL(A,B)	Matrix multiplication AB
MINVAL(A)	Minimum value in array A
MAXVAL(A)	Maximum value in array A
PRODUCT(A)	Product of values in array A
SUM(A)	Sum of values in array A
MAXLOC(A)	Location of maximum value in A
MINLOC(A)	Location of minimum value in A

The logical expression is evaluated, and if it is true, the block of statements is executed. Control then returns to the DO WHILE statement, and the loop continues to be executed as long as the logical expression is true.

B-5   POINTER VARIABLES

The inclusion of pointers in Fortran 90 allows us to generate and use a very important class of data structures—linked data structures. Because of the term *pointer,* we tend to think of a pointer variable as pointing to another variable. However, a pointer is more general and thus should be considered to be a name that is associated with a data object, which might be a variable, or a data structure, or a subset of a data structure. A pointer can also be considered to be an alias to a data object.

A variable that is to be used as a pointer must be defined with the pointer attribute and must be the same type as the object with which it is to be associated. Similarly, an object aliased by a pointer must be given a target attribute, as in the following statements:

```
REAL, TARGET :: X
REAL, POINTER :: TOP
```

The statement to associate the pointer TOP with X is the following:

```
TOP => X
```

A data object can have multiple pointers associated with it.

SUMMARY

This appendix can only begin to describe the powerful new features in this new standard. Fortran 90 compilers are currently available for many computers. The additional computational capabilities, the new data structures, and the new control structures will make it even easier to write readable programs to solve many different types of engineering problems. Since the new standard also incorporates the FORTRAN 77 standard, old programs will still execute properly while new programs will be easier to write and simpler to debug. The document that describes the complete language is entitled American National Standard Programming Language Fortran 90, ANSI X3.198-1991. It is identical to ISO/IEC 1539:1991, which is the International Fortran Standard.

Answers to Try It! Exercises

Try It! (page 13)

1. invalid (decimal point)
2. valid
3. valid
4. invalid (too long)
5. invalid character (__)
6. invalid character (__)
7. invalid characters (parentheses)
8. invalid (starts with a digit)
9. valid
10. invalid character (%)
11. not the same (15.7, 0.00157)
12. not the same (−1.7, 0.17)
13. same
14. not the same (0.005, 0.0000005)
15. not the same (0.899, 8990)
16. not the same (−0.044, 0.044)

Try It! (page 20)

1. Y = 12.5
2. X = 4.1
3. RESULT = −13

Try It! (page 21)

1. M = SQRT(X**2 + Y**2)
2. U = (U + V)/(1.0 + (U*V/C**2))
3. Y = YO*EXP(−A*T)*COS(2.0*PI*F*T)
4. T = ((5.0/9.0)*(TF − 32.0)) + 273.15
5. $PE = \dfrac{-G \cdot ME \cdot M}{R}$

6. $DF = E \cdot DA \cdot \cos \theta$

7. $AV = \dfrac{x2 - x1}{T2 - T1}$

8. $CA = \dfrac{4\pi^2 \cdot R}{T^2}$

9. $DIST = V \cdot TIME + \dfrac{ACC \cdot TIME^2}{2}$

10. $P = PO \cdot e^{\frac{-M \cdot G \cdot X \cdot TK}{R}}$

Try It! (page 34)

1. $X_b = bbb-27.6_b DEGREES$
2. $DISTANCE_b = bb0.287E_b05bbbbbVELOCITY_b = b-2.60$

Try It! (page 45)

1. $bb3.50bbbbbbbbbbbbb*******b0.0020$
2. $TIME_b = bb3.50bbbbbRESPONSE_b1_b = bb0.18E+03$
   $TIME_b = bb3.50bbbbbRESPONSE_b2_b = bb0.20E-02$
3. $EXPERIMENT_b RESULTS$

   $TIME_{bb}RESPONSE_b1_{bb}RESPONSE_b2$
   $3.50bbbbb178.800bbbbbbb0.002$
4. XX.XXX$_b$XX.XXX$_b$XX.XXX$_b$XX.XXX
5. XXX.XX

   XXX.XX

   XXX.XX

   XXX.XX
6. X.XX

   XX.X

   X.XX

   XX.X
7. XX.XXXXX.XXX

   XX.XXXXX.XX

Try It! (page 62)

1. false
2. false
3. false
4. true
5. true
6. false
7. false
8. true
9. IF (TIME.GT.5.0) TIME = TIME + 0.5
10. IF (SQRT(POLY).GE.8.0) PRINT*, 'POLY = ', POLY
11. DIFF = ABS(VOLT1 - VOLT2)
    IF (DIFF.LT.6.0) THEN
        PRINT*, 'VOLT1 = ', VOLT1

```
       PRINT*, 'VOLT2 = ', VOLT2
    END IF
12. IF (ABS(DEN).LT.0.005) PRINT*, 'DENOMINATOR IS TOO SMALL'
13. LNVAL = LOG(X**2)
    IF (LNVAL.GE.3.0) THEN
       TIME = 0.0
       SUM = SUM + X
    END IF
14. IF ((DIST.LT.50.0).OR.(TIME.GT.10.0)) THEN
       TIME = TIME + 1
    ELSE
       TIME = TIME + 0.5
    END IF
15. IF (DIST.GE.50.0) THEN
       TIME = TIME + 2.0
       PRINT*, 'DISTANCE > 50.0'
    END IF
16. IF (DIST.GT.100.0) THEN
       TIME = TIME + 2.0
    ELSE IF (DIST.GT.50.0) THEN
       TIME = TIME + 1.0
    ELSE
       TIME = TIME + 0.5
    END IF
```

Try It! (page 79)

1. 10 times
2. 9 times
3. 13 times
4. 7 times
5. 9 times
6. 91 times
7. COUNT = 8
8. COUNT = 15
9. COUNT = -16
10. COUNT = 18
11. COUNT = -1

Try It! (page 88)

1. COUNT = 101
2. COUNT = 36
3. COUNT = 10

Try It! (page 97)

1. TIME = 0.0
   TEMP = 86.3

2. TIME = 0.0
   TEMP = 0.5
3. TIME1 = 0.0
   TEMP1 = 86.3
   TIME2 = 0.5
   TEMP2 = 93.5
4. TIME1 = 0.0
   TEMP1 = 86.3
   TIME2 = 0.5
   TEMP2 = 93.5
5. TIME1 = 0.0
   TEMP1 = 0.5
   TIME2 = 1.0
   TEMP2 = 1.5
6. TIME1 = 0.0
   TEMP1 = 0.5
   TIME2 = 86.3
   TEMP2 = 93.5

Try It! (page 104)

1. works
2. works
3. would not work—time must be −99.0
4. would not work—the program will try to read a value for the variable ALT, causing an execution error
5. works

Try It! (page 123)

1. M(b1)b=bbb10
   M(b2)b=bbbb9
   M(b3)b=bbbb8
   M(b4)b=bbbb7
   M(b5)b=bbbb6
   M(b6)b=bbbb5
   M(b7)b=bbbb4
   M(b8)b=bbbb3
   M(b9)b=bbbb2
   M(10)b=bbbb1
2. bb-3bbb0bbb5bb12bb21bb32bb45bb60
3. TIMEbb1b=bb3.00
   TIMEbb5b=bb5.00
   TIMEbb9b=bb7.00
   TIMEb13b=bb9.00
   TIMEb17b=b11.00

Try It! (page 138)

1.

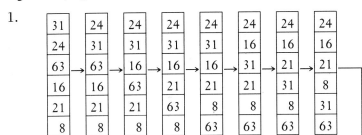

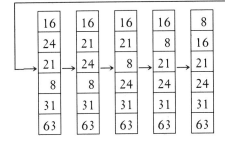

Try It! (page 143)

1.

4	6	8	10
6	8	10	12
8	10	12	14
10	12	14	16
12	14	16	18

bbb¹²bbb¹⁴bbb¹⁶bbb¹⁸

2.

1	2	0
2	3	1
3	4	2

bbb¹bbb²bbb⁰
bbb²bbb³bbb¹
bbb³bbb⁴bbb²

3.

11.5	14.5	17.5
13.0	16.0	19.0

b¹¹·⁵b¹⁴·⁵b¹⁷·⁵
b¹³·⁰b¹⁶·⁰b¹⁹·⁰

Try It! (page 160)

1. AREA(SIDE) = SIDE**2
2. AREA(BASE,HEIGHT) = BASE*HEIGHT
3. AREA(BASE, HT1, HT2) = 0.5*BASE*(HT1 + HT2)

Try It! (page 171)

1. S = 5.8
2. S = 2.9
3. S = 5.3
4. S = 12.2

Try It! (page 195)

1. NEW VALUES OF K ARE:
   bbbb^{1}bbbb^{2}bbbb^{3}bbbb^{0}bbbb^1
   bbbb^{2}bbbb^{3}bbbb^{0}bbbb^{1}bbbb^2
2. NEW VALUES OF K ARE:
   bbbb^{1}bbbb^{2}bbbb^{3}bbbb^{0}bbbb^1
   bbbb^{2}bbbb^{3}bbbb^{0}bbbb^{9}bbb^{1}0

Try It! (page 220)

1. 'TEN$_b$TOP$_b$ENGINEERING$_b$'
2. 'TEN$_b$TOP$_b$'
3. 'ENGINEERING'
4. 'A'
5. 'ENGINEERING$_b$ACHIEVEMENTS$_{bbb}$'
6. 'TEN$_b$TOP$_b$'
7. 'TEN$_b$TOP$_b$ENGINEERING$_b$ACHIEVEMENTS$_{bbb}$'
8. 'TEN$_b$ACHIEVEMENTS$_{bbb}$'
9. 'TOP$_b$TEN'
10. 'ENGINEERS'$_b$ACHIEVEMENTS$_{bbb}$'
11. LASER$_b$
12. FIBER$_b$
13. CADCAM
14. '''$_{bbb}$
15. $_{bb}$12.4
16. GENETI

Try It! (page 229)

1. 0.75D+00
2. 1.3D+00
3. 1.0D+00/9.0D+00
4. 5.0D+00/6.0D+00
5. $-$10.5D+00
6. 3.0D+00
7. $_{bbb}$0.78692D$-$02
8. $_{bbb}$0.787D$-$02
9. ************

Try It! (page 232)

1. CX = 0.0 + 3.0i
2. CX = $-$3.0 + 3.0i
3. CX = 2.0 + 1.0i
4. CX = $-$2.0 + 0.0i
5. CX = 3.5 $-$ 1.5i
6. CX = 1.41421 + 0.0i
7. $_{bb}$2.0+$_b$$-1.0_b$i
8. ($_{bb}$2.00,$_b$$-$1.00)

Answers to Selected Problems

Answers that contain FORTRAN statements are not usually unique. Although these answers represent good solutions to the problems, they are not necessarily the only valid solutions.

Chapter 2

1.

```
*------------------------------------------------------------------*
      PROGRAM CONVRT
*
*   This program converts kilowatt-hours to calories.
*
      REAL KWH, JOULES, CALRS
*
      PRINT*, 'ENTER ENERGY IN KILOWATT HOURS'
      READ*, KWH
*
      JOULES = 3.6E+06*KWH
      CALRS = JOULES/4.19
*
      PRINT 5, KWH, CALRS
    5 FORMAT (1X,F6.2,' KILOWATT-HOURS = ',E9.2,' CALORIES')
*
      END
*------------------------------------------------------------------*
```

Chapter 3

7.

```
*------------------------------------------------------------------*
      PROGRAM TABLE
*
```

```
*  This program prints a table of values for F where F = XY - 1
*  for a given set of X and Y values.
*
      INTEGER I
      REAL F, X, Y
*
      PRINT*, '   X  ', '   Y  ', '   F  '
      Y = 0.5
      DO 10 I = 1, 9
         X = REAL(I)
         F = X*Y - 1.0
         PRINT 5, X, Y, F
   5     FORMAT (1X,2X,F3.1,4X,F4.2,3X,F4.1)
         Y = Y + 0.25
  10 CONTINUE
*
      END
*----------------------------------------------------------------*
```

Chapter 4

2.
```
*----------------------------------------------------------------*
      PROGRAM PULSE
*
*  This program generates a sonar signal.
*
      INTEGER NUMBER, K
      REAL E, PD, FREQ, A, PERIOD, T, PI, TIME, SIGNAL
*
      OPEN (UNIT=10,FILE='SONAR',STATUS'NEW')
*
      PRINT*, 'ENTER TRANSMITTED ENERGY IN JOULES'
      READ*, E
      PRINT*, 'ENTER PULSE DURATION IN SECONDS'
      READ*, PD
      PRINT*, 'ENTER COSINE FREQUENCY IN HERTZ'
      READ*, FREQ
*
      A = SQRT(2.0*E/PD)
      PERIOD = 1.0/FREQ
      T = PERIOD/10.0
      NUMBER = NINT(PD/T)
      PI = 3.141593
*
      DO 10 K=0,NUMBER-1
         TIME = REAL(K)*T
         SIGNAL = A*COS(2.0*PI*FREQ*TIME)
         WRITE (10,*) TIME, SIGNAL
  10  CONTINUE
*
      TIME = -99.0
      SIGNAL = -99.0
      WRITE (10,*) TIME, SIGNAL
```

```
*
      END
*-------------------------------------------------------------------*
```

Chapter 5

3.

```
*-------------------------------------------------------------------*
      PROGRAM PWRPLT
*
*  This program computes and prints a composite report
*  summarizing several weeks of power plant data.
*
      INTEGER POWER(20,7), MIN, TOTAL, COUNT, I, J, N
      REAL AVE
      DATA TOTAL, COUNT /0, 0/
*
      PRINT*, 'ENTER NUMBER OF WEEKS FOR ANALYSIS'
      READ*, N
*
      OPEN (UNIT=12,FILE='PLANT',STATUS='OLD')
      DO 5 I=1,N
        READ (12,*) (POWER(I,J), J=1,7)
    5 CONTINUE
*
      MIN = POWER(1,1)
      DO 15 I=1,N
        DO 10 J=1,7
          TOTAL = TOTAL + POWER(I,J)
          IF (POWER(I,J).LT.MIN) MIN = POWER(I,J)
   10     CONTINUE
   15 CONTINUE
      AVE = REAL(TOTAL)/REAL(N*7)
*
      DO 25 I=1,N
        DO 20 J=1,7
          IF (POWER(I,J).GT.AVE) COUNT = COUNT + 1
   20     CONTINUE
   25 CONTINUE
*
      PRINT 30
   30 FORMAT (1X,15X,'COMPOSITE INFORMATION')
      PRINT 35, AVE
   35 FORMAT (1X,'AVERAGE DAILY POWER OUTPUT = ',F5.1,
     +        ' MEGAWATTS')
      PRINT 40, COUNT
   40 FORMAT (1X,'NUMBER OF DAYS WITH GREATER THAN ',
     +        'AVERAGE POWER OUTPUT = ',I2)
      PRINT 45
   45 FORMAT (1X,'DAYS(S) WITH MINIMUM POWER OUTPUT:')
      DO 60 I=1,N
        DO 55 J=1,7
          IF (POWER(I,J).EQ.MIN) PRINT 50, I, J
   50     FORMAT (1X,12X,'WEEK ',I2,'   DAY ',I2)
```

```
      55    CONTINUE
      60 CONTINUE
*
         END
*------------------------------------------------------------------*
```

Chapter 6

14.

```
*------------------------------------------------------------------*
         INTEGER FUNCTION INVERT(NUM)
*
*  This function reverses the digits in a two-digit number.
*
         INTEGER NUM, DIGIT1, DIGIT2
*
         DIGIT1 = NUM/10
         DIGIT2 = MOD(NUM,10)
         INVERT = DIGIT2*10 + DIGIT1
*
         RETURN
         END
*------------------------------------------------------------------*
```

Chapter 7

1.

```
*------------------------------------------------------------------*
         PROGRAM SIGGEN
*
*  This program generates a signal composed of a sine wave
*  plus random noise.
*
         INTEGER SEED, I, N
         REAL PI, T, NOISE, X
         DATA PI, T /3.141593, 0.0/
*
         PRINT*, 'ENTER A POSITIVE INTEGER SEED: '
         READ*, SEED
         OPEN (UNIT=15,FILE='SIGNAL',STATUS='NEW')
*
         PRINT*, 'ENTER NUMBER OF DATA POINTS TO GENERATE'
         READ*, N
*
         DO 10 I=1,N
            CALL RANDOM (SEED, NOISE)
            X = 2*SIN(2*PI*T) + NOISE
            WRITE (15,*) T, X
            T = T + 0.01
      10 CONTINUE
*
         END
*------------------------------------------------------------------*
```

(no changes in subroutine RANDOM)
--

Chapter 8

4.
--

```
      PROGRAM WEIGHT
*
*  This program reads character strings containing amino acids
*  from large protein molecules, and computes the molecular
*  weights.
*
      INTEGER K, MW(20), N, J, BLNK, NCHAR, AMNUM, SUM,
     +        START, AMWT, COUNT
      CHARACTER*3 REF(20)
      CHARACTER*150 PROTN
      LOGICAL ERROR
*
      OPEN (UNIT=9,FILE='AMINO',STATUS='OLD')
      DO 10 K=1,20
         READ(9,*) REF(K), MW(K)
   10 CONTINUE
*
      OPEN (UNIT=10,FILE='PROTEIN',STATUS='OLD')
      READ(10,*) N
*
      DO 30 J=1,N
*
         READ(10,*) PROTN
         BLNK = INDEX(PROTN,' ')
         NCHAR = BLNK - 1
         AMNUM = NCHAR/3
         SUM = 0
*
         IF (MOD(NCHAR,3).NE.0) THEN
*
            PRINT*, 'LENGTH ERROR IN PROTEIN ', J
            PRINT*, PROTN(:BLNK)
            PRINT*
*
         ELSE
*
```

```
                              ERROR = .FALSE.
                              COUNT = 0
                              DO 20 K=1,AMNUM
                                 START = (K-1)*3 + 1
                                 CALL MWT(REF,MW,PROTN(START:START+2),AMWT)
                                 IF (AMWT.NE.0) THEN
                                    SUM = SUM + AMWT
                                    COUNT = COUNT + 1
                                 ELSE
                                    PRINT*, 'ERROR IN AMINO ', K, ' PROTEIN ', J
                                    ERROR = .TRUE.
                                 END IF
         20                   CONTINUE
                              IF (.NOT.ERROR) THEN
                                 PRINT 15, PROTN(:BLNK), SUM
         15                   FORMAT (1X,'PROTEIN: ',A/
                +                      1X,'MOLECULAR WEIGHT:',I8/)
                                 PRINT 25, COUNT
         25                   FORMAT (1X,'NUMBER OF AMINO ACIDS IS ',I4/)
                              ELSE
                                 PRINT*, PROTN(:BLNK)
                                 PRINT*
                              END IF
*
              END IF
*
   30 CONTINUE
*
      END
*-------------------------------------------------------------------*
      SUBROUTINE MWT(REF,MW,STRING,WEIGHT)
*
* This subroutine receives arrays containing the character
* references and molecular weights for amino acids. It uses
* these arrays to determine if an input string is an amino acid
* and, if so, returns the molecular weight of the amino acid;
* otherwise zero is returned.
*
      INTEGER MW(20), WEIGHT, K
      CHARACTER*3 REF(20), STRING
*
      WEIGHT = 0
      DO 10 K=1,20
         IF (STRING.EQ.REF(K)) WEIGHT = MW(K)
   10 CONTINUE
*
      RETURN
      END
*-------------------------------------------------------------------*
```

Index